AF771385

CONTENTS

CAN FEDERAL AGENCIES FUNCTION IN THE WAKE OF A DISASTER? A STATUS REPORT ON FEDERAL AGENCIES' CONTINUITY OF OPERATIONS PLANS

HEARING

BEFORE THE

COMMITTEE ON GOVERNMENT REFORM

HOUSE OF REPRESENTATIVES

ONE HUNDRED EIGHTH CONGRESS

SECOND SESSION

APRIL 22, 2004

Serial No. 108–184

Printed for the use of the Committee on Government Reform

Available via the World Wide Web: http://www.gpo.gov/congress/house
http://www.house.gov/reform

U.S. GOVERNMENT PRINTING OFFICE

95–423 PDF WASHINGTON : 2004

For sale by the Superintendent of Documents, U.S. Government Printing Office
Internet: bookstore.gpo.gov Phone: toll free (866) 512–1800; DC area (202) 512–1800
Fax: (202) 512–2250 Mail: Stop SSOP, Washington, DC 20402–0001

CAN FEDERAL AGENCIES FUNCTION IN THE WAKE OF A DISASTER? A STATUS REPORT ON FEDERAL AGENCIES' CONTINUITY OF OPERATIONS PLANS

THURSDAY, APRIL 22, 2004

HOUSE OF REPRESENTATIVES,
COMMITTEE ON GOVERNMENT REFORM,
Washington, DC.

The committee met, pursuant to notice, at 10:07 a.m., in room 2154, Rayburn House Office Building, Hon. Tom Davis (chairman of the committee) presiding.

Present: Representatives Tom Davis, Ose, Jo Ann Davis, Blackburn, Maloney, Cummings, Tierney, Watson, Van Hollen, Ruppersberger and Norton.

Staff present: David Marin, deputy staff director/director of communications; Anne Marie Turner and John Hunter, counsels; Robert Borden, counsel/parliamentarian; Drew Crockett, deputy director of communications; John Cuaderes, senior professional staff member; Teresa Austin, chief clerk; Brien Beattie, deputy clerk; Corinne Zaccagnini, chief information officer; Robert White, press secretary; Michael Yeager, minority deputy chief counsel; Earley Green, minority chief clerk; and Jean Gosa, minority assistant clerk.

Chairman TOM DAVIS. Good morning. A quorum being present, the Committee on Government Reform will come to order. I would like to welcome everyone to today's hearing on the status of the Federal Government's continuity of operations plans.

Today on the House floor we are considering legislation laying out the framework for how Congress would continue operating in the event of a catastrophe. That's important. But let's be honest. The real, tangible, day-to-day work of the Federal Government doesn't happen here. It happens at agencies spread across the Nation, and ensuring their continued operation in the wake of a devastating tragedy should be considered every bit as important.

Continuity of Federal Government operations planning became essential during the cold war, to protect the continuity of government in the event of a nuclear attack. COOP planning has attracted renewed significance after the terrorist attacks of September 11. Through a Presidential Decision Directive and a Federal Preparedness Circular, Federal agencies are required to develop viable continuity of operations plans for ensuring the continuity of essential operations in emergency situations. Although it is a classified document, PDD 67 reportedly also designates the Federal

Emergency Management Association [FEMA], as the executive agency for formulating guidance on executive departments' COOP plans, and coordinating and assessing their capabilities. In July 1999, FEMA issued Federal Preparedness Circular 65, FPC 65, which confirms its coordinating agency role, contains criteria for agencies to develop their plans, and designates the timelines for submission of agency plans.

Because of the critical nature of the ongoing threat of emergencies, including terrorist attacks, severe weather, and individual building emergencies, this committee requested the GAO to evaluate contingency plans of several Federal agencies and review FEMA's oversight of those agency COOP plans. And in February 2004, GAO issued a report that found a wide variance of essential functions identified by individual agencies. GAO attributed this lack of uniformity to several factors: lack of specificity about criteria to identify essential functions in FPC 65; lack of review by FEMA of essential functions during assessment of COOP planning; lack of testing or exercises by FEMA to confirm the identification of essential functions by agencies.

To remedy these shortcomings, GAO recommends that the Secretary of the Department of Homeland Security direct the Under Secretary for Emergency Preparedness and Response to ensure that agencies develop COOP plans by May 1, 2004 and correct deficiencies in individual plans. In addition, GAO recommends that the Under Secretary be directed to conduct assessments of COOP plans that include independent verification of agency information, agencies' essential functions and their interdependencies with other activities.

The committee is concerned about the seeming lack of progress we have made in the area of Federal continuity of operations. If September 11 was a wakeup call, then we haven't fully heeded the message when it comes to our planning. Although some progress has been made, and I commend Under Secretary Brown for his leadership on this, we still have a ways to go. We must do everything possible to address the COOP inconsistencies that exist across the board. Identifying and prioritizing essential functions with 100 percent compliance and accuracy is a must. Even if agencies can accomplish this, they still must be able to identify their key staffing requirements, lines of succession, resources needed, and what mission-critical systems and data must be protected and, in many cases, be redundant.

Continuity of operations means more than keeping your Web site up and running. What's really called for is a wholistic approach, one that factors in people, places and things. What is really needed is agility, because FEMA's role in COOP oversight is key for agency success.

The committee will hear FEMA's assessment of the individual agency plans. The committee will also assess FEMA's efforts to ensure that the COOP directives are carried out by each agency. This will include steps FEMA is taking to assess each of the executive agencies' COOP plans, what interaction FEMA has had and plans to have with those agencies about deficiencies in those plans, what steps FEMA will take to ensure agency compliance, and FEMA's assessment of the adequacy of Federal Preparedness Circular 65, and steps it has taken to overcome any deficiencies.

The committee will also hear from GAO about its assessment of COOP planning and its recommendations for improvement and will also hear how the private sector deals with this issue.

Finally, the committee has asked GAO to continue to monitor Federal COOP planning to ensure that agencies are in compliance with the latest executive and congressional guidance. The committee expects to get an annual scorecard from GAO outlining how agencies are performing with regard to the many facets of COOP. This is an important issue and we'll be very aggressive on our oversight.

We have three impressive witnesses before us to help us understand the current and future state of Federal continuity of operations planning, the expected problems and what we can look forward to in ways of improvement. First we will hear from the General Accounting Office, followed by the Department of Homeland Security, and finally we will hear from AT&T which has a mature COOP plan in place.

I want to thank all of our witnesses for appearing before the committee and I look forward to hearing their testimony.

[The prepared statement of Chairman Tom Davis follows:]

Opening Statement
Chairman Tom Davis
Committee on Government Reform
"Can Federal Agencies Function in the Wake of a Disaster?
A Status Report on Federal Agencies' Continuity of Operations Plans"
April 22, 2004

Good morning. A quorum being present, the Committee on Government Reform will come to order. I would like to welcome everyone to today's hearing on the status of the Federal government's continuity of operations plans. Today on the House floor, we are considering legislation laying out the framework for how Congress would continue operating in the event of a catastrophe. That is important, but let's be honest: the real, tangible, day-to-day work of the Federal government doesn't happen here. It happens at agencies spread across our nation, and ensuring their continued operations in the wake of a devastating tragedy should be considered every bit as important.

Continuity of Federal government operations planning became essential during the Cold War to protect the continuity of government in the event of a nuclear attack. COOP planning has attracted renewed significance after the terrorist attacks of September 11. Through a Presidential Decision Directive and a Federal Preparedness Circular, Federal agencies are required to develop viable COOP plans for ensuring the continuity of essential operations in emergency situations. Although it is a classified document, PDD 67 reportedly also designates the Federal Emergency Management Agency (FEMA) as the executive agent for formulating guidance on executive departments' COOP plans, and coordinating and assessing their capabilities. In July 1999, FEMA issued Federal Preparedness Circular (FPC) 65, which confirms its coordinating agency role, contains criteria for agencies to develop their plans, and designates timelines for submission of agency plans.

Because of the critical nature of the ongoing threat of emergencies, including terrorist attacks, severe weather, and individual building emergencies, this Committee requested the General Accounting Office (GAO) to evaluate contingency plans of several Federal agencies and review FEMA's oversight of those agencies' COOP plans. In February 2004, GAO issued a report that found a wide variance of essential functions identified by individual agencies. GAO attributed this lack of uniformity to several factors:

- lack of specificity of criteria to identify essential functions in FPC 65;
- lack of review by FEMA of essential functions during assessment of COOP planning; and
- lack of testing or exercises by FEMA to confirm the identification of essential functions by agencies.

To remedy these shortcomings, GAO recommends that the Secretary of DHS direct the Under Secretary for Emergency Preparedness and Response to ensure that agencies develop COOP plans by May 1, 2004 and correct deficiencies in individual COOP plans. In addition, GAO recommends that the Under Secretary be directed to conduct assessments of COOP plans that include independent verification of agency information, agencies' essential functions and their interdependencies with other activities.

The Committee is concerned about the seeming lack of progress we have made in the area of Federal continuity of operations. If 9-11was the wake up call, then we haven't fully heeded the message when it comes to COOP. Although some progress has been made, and I commend Under Secretary Brown for his leadership on this, we still have a ways to go. We must do everything possible to address the COOP inconsistencies that exist across the board. Identifying and prioritizing essential functions with 100% compliance and accuracy is a must. Even if agencies can accomplish this, they still must be able to identify their key staffing requirements, lines of succession, resources needed and what mission-critical systems and data must be protected and, in many cases, be redundant.

Continuity of operations means more than keeping your web site up and running. What's really called for is a holistic approach, one that factors in people, places and things. What's really needed is agility.

Because FEMA's role in COOP oversight is key for agency success, the Committee will hear FEMA's assessment of the individual agencies' COOP plans. The Committee will also assess FEMA's efforts to ensure that the COOP directives are carried out by each agency. This will include steps FEMA is taking to assess each of the executive agencies' COOP plans, what interaction FEMA has had and plans to have with those agencies about deficiencies in those plans, what steps FEMA will take to ensure agency compliance, and FEMA's assessment of the adequacy of Federal Preparedness Circular 65 and steps it is taking to overcome any deficiencies. The Committee will also hear from GAO about its assessment of COOP planning and its recommendations for improvement.

Finally, the Committee has asked GAO to continue to monitor federal COOP planning to ensure that agencies are in compliance with the latest executive and congressional guidance. The Committee expects to get an annual "scorecard" from GAO outlining how agencies are performing with regard to the many facets of COOP. This is an important issue and we will be very aggressive on our oversight.

We have three impressive witnesses before us to help us understand the current and future state of Federal continuity of operations planning, the expected problems and what we can look forward to in ways of improvements. First we will hear from the General Accounting Office, followed by the Department of Homeland Security, and finally we will hear from AT&T, which has a mature COOP plan in place. I would like to thank all of our witnesses for appearing before the Committee, and I look forward to their testimony.

Chairman TOM DAVIS. Are there any other Members who wish to make opening statements at this point?

Ms. Watson.

Ms. WATSON. Thank you very much Mr. Chairman.

In the event of a crisis, the American people immediately turn to the Federal Government to provide basic services, stability and direction. But we now have learned from the GAO that many Federal agencies are woefully unprepared to continue functioning in the wake of a catastrophe. It is distressing to know that in the wake of an attack on America, the horror of the initial attack might be compounded by the mayhem of a government that cannot coordinate basic services. We need to fix this.

And I think all of us have it indelible in our minds where we were and what we were doing on September 11, myself included, right here in this Capitol. And we knew not where to go. We were running around like ants all over the place. We knew not where to gather. I had to seek out directions. And we have to be sure that we have these plans in place.

But this is only the tip of the iceberg. Even beyond this, what is not addressed in this report or in this hearing is continuity of operations at the State or at the local level. I bring this issue up, Mr. Chairman, not to confuse the issue in this hearing, which I understand focuses solely on the continuity of operations and planning in the Federal executive branch, but rather simply to illustrate the scope of the problem that we face. Even once we get this problem sorted out at the Federal level, we must ensure our States and our local governments that they are prepared. Here we sit, 2½ years after facing the mortal threat of September 11, and we still cannot be assured that we are prepared to provide essential government services in the wake of a disaster.

My colleagues and I want some answers. And I ask the witnesses from FEMA, please tell us what you need to tell us, and we will do our best to see that you get it. But we need to hear from you, and we need to know what your plans are for real progress and real answers, and on how you prepare to fix it. And I'm sure you will find this Congress very supportive.

Thank you, Mr. Chairman.

Chairman TOM DAVIS. Thank you.

Any other Members wish to make opening statements? If not, we will move to our first witness, Linda Koontz, the Director of Information Management Issues of the General Accounting Office, no stranger to this committee. As you know it's the policy of the committee that all witnesses be sworn in before they testify. So, Linda, if you'd rise with me and raise your right hand.

[Witness sworn.]

Chairman TOM DAVIS. For the record, note we have—two of your aides behind you also sworn in. Please proceed with your testimony. You know the rules. We have the buttons, the lights out here, 5 minutes and try to sum up. And thank you for being with us again.

STATEMENT OF LINDA D. KOONTZ, DIRECTOR, INFORMATION MANAGEMENT ISSUES, U.S. GENERAL ACCOUNTING OFFICE

Ms. KOONTZ. Thank you, Mr. Chairman and members of the committee. I appreciate the opportunity to participate in the committee's hearing on Federal continuity of operations planning. As you know, events such as terrorist attacks, severe weather, or building-level emergencies can disrupt the delivery of essential government services. To minimize the risk of disruption, Federal agencies are required to develop plans for ensuring the continuity of essential services in emergency situations.

The Federal Emergency Management Agency, now part of the Department of Homeland Security, was designated executive agent for continuity of operations planning and issued guidance in July 1999. This guidance states that in order to have a viable continuity of operations capability, agencies should identify their essential functions. Identifying essential functions is the first of eight elements of a viable capability and provides the basis for subsequent planning steps.

Mr. Chairman, at your request, we assessed department and agency-level continuity of operations plans at 23 major Federal agencies and reported the results to you in February. In summary, we found that, first, three departments did not have plans in place as of October 1, 2002. Second, our assessment raised serious questions about the adequacy of the essential functions identified. Specifically, we found that 29 of the 34 plans that we reviewed identified at least one essential function. However, these functions varied widely in number from 3 to 399, and included many that appeared to be of secondary importance.

At the same time, the plans omitted many programs that OMB had previously identified as having a high impact on the public. Agencies did not list among their essential functions 20 of the 38 high-impact programs that have been previously identified. For example, one department included, "provided speeches and articles for the Secretary and Deputy Secretary," among its essential functions, but did not include 9 of 10 high-impact programs. In addition, although many agency functions rely on the availability of resources or functions controlled by another organization, more than three-fourths of the plans did not fully identify such dependencies.

Third, none of the agencies provided documentation sufficient to show that they were complying with all aspects of FEMA's guidance.

In our view, a number of factors contributed to these government-wide shortcomings. FEMA's planning guidance does not provide specific criteria for identifying essential functions, nor does it address interdependencies. In addition, while FEMA conducted an assessment of agency compliance with the guidance in 1999, it has not conducted oversight that is sufficiently regular and extensive to ensure that agencies correct deficiencies identified. Further, in its assessment, FEMA did not include a review of essential functions. Finally, FEMA did not conduct tests or exercises to confirm that the identified essential functions were correct.

In discussing our report, FEMA officials, while maintaining that the government is prepared for an emergency, acknowledged that improvements could be made. These officials told us that they plan to conduct a government-wide exercise next month, improve oversight by providing more detailed planning guidance, and develop a system to collect data from agencies on their readiness. However, these officials have not yet determined how they will verify the agency-reported data, assess the essential function and interdependencies identified, or use the data to conduct regular oversight. In our report, we made several recommendations to address these shortcomings.

In summary, Mr. Chairman, while most of the agencies reviewed had continuity of operation plans in place, those plans exhibited weaknesses in the form of widely varying determinations about what functions are essential, and inconsistent compliance with guidance that defines a viable continuity of operations capability. Until these weaknesses are addressed, agencies are likely to continue to base their plans on ill-defined assumptions that may limit the utility of the resulting plans, and, as a result, risk experiencing difficulties in delivering key services to citizens in the aftermath of an emergency.

Mr. Chairman, that concludes my statement. I would be happy to answer any questions that you might have.

Chairman TOM DAVIS. Thank you.

[The prepared statement of Ms. Koontz follows:]

United States General Accounting Office

GAO

Testimony Before the Committee on Government Reform, House of Representatives

For Release on Delivery
Expected at 10 a.m. EDT
on Thursday
April 22, 2004

CONTINUITY OF OPERATIONS

Improved Planning Needed to Ensure Delivery of Essential Services

Statement of Linda D. Koontz
Director, Information Management Issues

GAO
Accountability * Integrity * Reliability

GAO-04-638T

GAO
Accountability • Integrity • Reliability

Highlights

Highlights of GAO-04-638T, a testimony before the Committee on Government Reform, House of Representatives

April 2004

CONTINUITY OF OPERATIONS

Improved Planning Needed to Ensure Delivery of Essential Government Services

Why GAO Did This Study

To ensure that essential government services are available in emergencies—such as terrorist attacks, severe weather, or building-level emergencies—federal agencies are required to develop continuity of operations (COOP) plans. Responsibility for formulating guidance on these plans and for assessing executive branch COOP capabilities lies with the Federal Emergency Management Agency (FEMA), under the Department of Homeland Security. FEMA guidance, Federal Preparedness Circular (FPC) 65 (July 1999), identifies elements of a viable COOP capability, including the requirement that agencies identify their essential functions.

This statement summarizes the findings of a February 2004, GAO report, for which GAO was asked to determine the extent to which (1) major civilian executive branch agencies have identified their essential functions and (2) these agencies' COOP plans follow FEMA guidance.

What GAO Recommends

In a February 2004 report, GAO recommended that the Secretary of Homeland Security take steps to improve agency COOP plans and FEMA's process for assessing these plans. DHS agreed that improvements were needed in the COOP planning process, and that FEMA could do more to ensure such improvements were made.

www.gao.gov/cgi-bin/getrpt?GAO-04-638T.

To view the full product, including the scope and methodology, click on the link above. For more information, contact Linda Koontz at (202) 512-6240 or koontzl@gao.gov.

What GAO Found

Based on an assessment of 34 COOP plans against FEMA guidance, GAO found that most agencies' plans identified at least one function as essential. However, the functions identified in each plan varied widely in number—ranging from 3 to 399—and included functions that appeared to be of secondary importance, while at the same time omitting programs that had been previously defined as high-impact programs. For example, one department included "provide speeches and articles for the Secretary and Deputy Secretary," among its essential functions, but did not include 9 of 10 high-impact programs for which it was responsible. Several factors contributed to these shortcomings: FPC 65 did not provide specific criteria for identifying essential functions; FEMA did not review the essential functions identified when it assessed COOP planning; and it did not conduct tests or exercises to confirm that the essential functions were correctly identified. Unless agencies' essential functions are correctly and completely identified, their COOP plans may not effectively ensure that the most vital government services can be maintained in an emergency.

Although all but three of the agencies reviewed had developed and documented some of the elements of a viable COOP plan, none of the agencies could demonstrate that they were following all the guidance in FPC 65. As the figure shows, there is a wide variation in the number of agencies that addressed various elements identified in the guidance. A contributing cause for the deficiencies in agency COOP plans is the level of FEMA oversight. In 1999, FEMA conducted an assessment of agency compliance with FPC 65, but it has not conducted oversight that is sufficiently regular and extensive to ensure that agencies correct the deficiencies identified. Because the resulting COOP plans do not include all the elements of a viable plan as defined by FPC 65, agency efforts to provide services during an emergency could be impaired.

Elements That Were Included in Agency COOP Plans in Place as of October 1, 2002

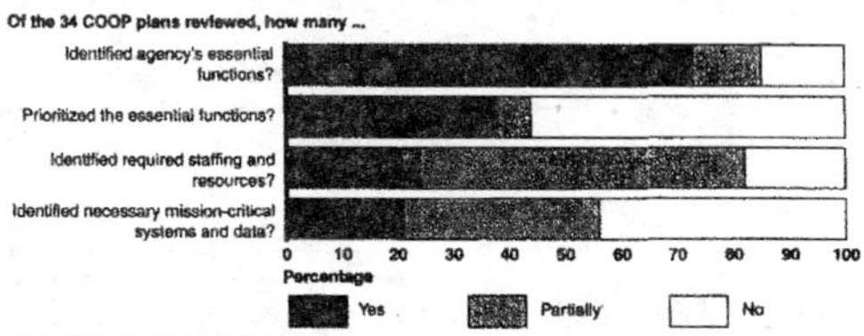

Source: GAO analysis of agency COOP plans.

Mr. Chairman and Members of the Committee:

I appreciate the opportunity to participate in the Committee's hearing on federal continuity of operations planning. As you know, events such as terrorist attacks, severe weather, or building-level emergencies can disrupt the delivery of essential government services. To minimize the risk of disruption, federal agencies are required to develop plans for ensuring the continuity of essential services in emergency situations. The Federal Emergency Management Agency (FEMA), which was designated executive agent for executive branch continuity of operations (COOP) planning, issued planning guidance in July 1999. This guidance, known as Federal Preparedness Circular (FPC) 65, states that in order to have a viable COOP capability, agencies should identify their essential functions. Identifying essential functions is the first of eight elements of a viable COOP capability, and provides the basis for subsequent planning steps.

At your request, we analyzed the continuity of operations plans in place at 20 major civilian departments and agencies[1] as of October 1, 2002. We reported the results of our analysis to you in February.[2] My remarks today will summarize those results. Specifically, I will discuss

- the extent to which agencies have identified their essential functions, and
- the extent to which their plans follow the guidance provided in FPC 65.

In conducting the analysis for our February report, we obtained and evaluated the headquarters contingency plans in place as of October 1, 2002, from 20 of the 23 largest civilian departments and agencies,

[1] Three of the selected major agencies did not have documented COOP plans in place as of October 1, 2002.

[2] U.S. General Accounting Office, *Continuity of Operations: Improved Planning Needed to Ensure Delivery of Essential Services*, GAO-04-160 (Washington, D.C.: Feb. 27, 2004.)

as well as the headquarters plans for 15 components of civilian cabinet-level departments, selected because they were responsible for programs previously deemed high impact by the Office of Management and Budget (OMB). (The major departments and agencies reviewed are listed in attachment I.)[3] We also reviewed supporting documentation and interviewed the agency officials responsible for developing these plans, obtained and analyzed FEMA guidance and documents describing its efforts to provide oversight and assessments of the federal continuity planning efforts, and interviewed FEMA officials to clarify the activities described in these documents. This testimony is based on previously published work, which was conducted in accordance with generally accepted government auditing standards, from April 2002 through January of this year.[4]

Results in Brief

Twenty-nine of the 34 COOP plans[5] that we reviewed identified at least one essential function. However, the functions identified in these plans varied widely in number—ranging from 3 to 399—and included functions that appeared to be of secondary importance. At the same time, the plans omitted many programs that OMB had previously identified as having a high impact on the public. Agencies did not list among their essential functions 20 of the 38 high-impact programs that had been identified at those agencies. For example, one department included "provide speeches and articles for the Secretary and Deputy Secretary" among its essential functions, but it did not include 9 of its 10 high-impact programs. In addition,

[3] Attachment II provides a list of the high-impact programs and the component agencies responsible for them. Attachment III identifies the 15 components whose COOP plans we reviewed and the high-impact programs for which they are responsible.

[4] We also reported on the human capital considerations relevant to COOP planning and implementation in U.S. General Accounting Office, *Human Capital: Opportunities to Improve Federal Continuity Planning Guidance*, GAO-04-384 (Washington, D.C.: Apr. 20, 2004).

[5] One COOP plan covered two components. As a result, the 34 COOP plans we reviewed covered 35 departments and agencies, including components.

although many agency functions rely on the availability of resources or functions controlled by another organization, more than three-fourths of the plans did not fully identify such dependencies. Several factors contributed to these governmentwide shortcomings: FPC 65 does not provide specific criteria for identifying essential functions, nor does it address interdependencies; FEMA did not review the essential functions identified in its assessments of COOP planning or follow up with agencies to determine whether they addressed previously identified weaknesses; and it did not conduct tests or exercises that could confirm that the identified essential functions were correct. Although the agency has begun efforts to develop additional guidance and conduct a governmentwide exercise, these actions have not yet been completed. Without better oversight, agencies are likely to continue to base their COOP plans on ill-defined assumptions that may limit the utility of the resulting plans.

While all but three of the agencies that we reviewed had developed and documented some elements of a COOP plan, none of the agencies provided documentation sufficient to show that they were following all the guidance in FPC 65. FEMA conducted an assessment of agency compliance with FPC 65 in 1999, but it has not conducted oversight that is sufficiently regular and extensive to ensure that agencies correct deficiencies identified. This limited level of oversight was a contributing cause for the deficiencies in agency COOP plans. FEMA officials told us that they plan to improve oversight by providing more detailed guidance and developing a system to collect data from agencies on their COOP readiness. However, the officials have not yet determined how they will verify the agency-reported data, assess the essential functions and interdependencies identified, or use the data to conduct regular oversight. If these shortcomings are not addressed, agency COOP plans may not be effective in ensuring that the most vital government services can be maintained in an emergency.

In our report, we made several recommendations to the Secretary of Homeland Security to enhance the ability of the federal government to provide essential services during emergencies. In response to a draft of our report, the Under Secretary for Emergency Preparedness and Response agreed that better planning is needed to ensure delivery of essential services, and that the department could do more to improve. He added that the department has begun to correct the identified deficiencies and stated that the federal government is currently poised to provide services in an emergency. Once the department assesses and independently verifies the status of agencies' plans, it will have convincing evidence to support such statements about readiness in the future.

Background

Federal operations and facilities have been disrupted by a range of events, including the terrorist attacks on September 11, 2001; the Oklahoma City bombing; localized shutdowns due to severe weather conditions, such as the closure of federal offices in Denver for 3 days in March 2003 due to snow; and building-level events, such as asbestos contamination at the Department of the Interior's headquarters. Such disruptions, particularly if prolonged, can lead to interruptions in essential government services. Prudent management, therefore, requires that federal agencies develop plans for dealing with emergency situations, including maintaining services, ensuring proper authority for government actions, and protecting vital assets.

Until relatively recently, continuity planning was generally the responsibility of individual agencies. In October 1998, Presidential Decision Directive (PDD) 67 identified FEMA—which is responsible for responding to, planning for, recovering from, and mitigating against disasters—as the executive agent for federal COOP planning across the federal executive branch. FEMA was an independent agency until March 2003, when it became part of the Department of Homeland Security, reporting to the Under Secretary for Emergency Preparedness and Response.

PDD 67 is a Top Secret document controlled by the National Security Council. FPC 65 states that PDD 67 made FEMA, as executive agent for COOP, responsible for

- formulating guidance for agencies to use in developing viable plans;
- coordinating interagency exercises and facilitating interagency coordination, as appropriate; and

- overseeing and assessing the status of COOP capabilities across the executive branch.

According to FEMA officials, PDD 67 also required that agencies have COOP plans in place by October 1999.

In July 1999, FEMA issued FPC 65 to assist agencies in meeting the October 1999 deadline. FPC 65 states that COOP planning should address any emergency or situation that could disrupt normal operations, including localized emergencies. FPC 65 also determined that COOP planning is based first on the identification of essential functions—that is, those functions that enable agencies to provide vital services, exercise civil authority, maintain safety, and sustain the economy during an emergency. FPC 65 gives no criteria for identifying essential functions beyond this definition.

Although FPC 65 gives no specific criteria for identifying essential functions, a logical starting point for this process would be to consider programs that had been previously identified as important. For example, in March 1999, as part of the efforts to address the Y2K computer problem,[6] the Director of OMB identified 42 programs with a high impact on the public:

- Of these 42 programs, 38 were the responsibility of the 23 major departments and agencies that we reviewed. (Attachment II provides a list of these 38 high-impact programs and the component agencies that are responsible for them.)
- Of these 23 major departments and agencies, 16 were responsible for at least one high-impact program; several were responsible for more than one.

Programs that were identified included weather service, disease monitoring and warnings, public housing, air traffic control, food stamps, and Social Security benefits. These programs, as well as the

[6] The need to ensure that computers would handle dates correctly in the year 2000 (Y2K) and beyond resulted in a governmentwide effort to identify mission-critical systems and high-impact programs supported by these systems.

others listed in attachment II, continue to perform important functions for the public.

The Y2K efforts to support such high-impact programs included requirements for COOP planning and the identification of interdependencies. Specifically, agencies were tasked with identifying partners integral to program delivery, testing data exchanges across partners, developing complementary business continuity and contingency plans, sharing key information on readiness with other partners and the public, and taking other steps to ensure that the agency's high-impact program would work in the event of an emergency.

In addition to requiring agencies to identify their essential functions, FPC 65 also defined an additional seven planning topics that make up a viable COOP capability. The guidance provided a general definition of each of the eight topics and identified several actions that should be completed to address each topic. Table 1 lists the eight topic areas covered in FPC 65 and provides an example of an action under each.

Table 1: Eight COOP Planning Topics Defined by FPC 65 and Examples of Actions

FPC 65 planning topic	Example of action (element of viable COOP plan)
Essential functions should be identified to provide the basis for COOP planning.	The agency should prioritize its essential functions.
Plans and procedures should be developed and documented to provide for continued performance of essential functions.	These plans should include a roster of personnel who can perform the essential functions.
Orders of succession should identify alternates to fill key positions in an emergency.	Succession lists should be developed for the agency head and other key positions.
Delegations of authority should identify the legal basis for officials to make decisions in emergencies.	Delegations should include the circumstances under which the authorities begin and end.
Alternate facilities should be able to support operations in a threat-free environment for up to 30 days.	These facilities should provide sufficient space and equipment to sustain the relocating organization.
Interoperable communications should provide voice and data communications with others inside and outside the organization.	The agency should be able to communicate with agency personnel, other agencies, critical customers, and the public.
Vital records should be identified and made readily available in an emergency.	Electronic and paper records should be identified and protected.
Tests, training, and exercises should occur regularly to demonstrate and improve agencies' COOP capabilities.	Individual and team training should be conducted annually.

Sources: FPC 65, FEMA.

Many COOP Plans Did Not Address Previously Identified Essential Functions or Interdependencies with Other Entities

The identification of essential functions is a prerequisite for COOP preparation because it establishes the parameters that drive the agency's efforts in all other planning topics. For example, FPC 65 directs agencies to identify alternative facilities, staff, and resources necessary to support continuation of their essential functions. The effectiveness of the plan as a whole and the implementation of all other elements depend on the performance of this step.

Of the 34 agency COOP plans that we reviewed, 29 plans included at least one function that was identified as essential. These agency-identified essential functions varied in number and scope. The number of functions identified in each plan ranged from 3 to 399. In addition, the apparent importance of the functions was not consistent. For example, a number of essential functions were of clear importance, such as

- "ensuring uninterrupted command, control, and leadership of the Department";
- "protecting critical facilities, systems, equipment and records"; and
- "continuing to pay the government's obligations."

Other identified functions appeared vague or of questionable importance:

- "provide speeches and articles for the Secretary and Deputy Secretary";
- "schedule all activities of the Secretary"; and
- "review fiscal and programmatic integrity and efficiency of Departmental activities."

In contrast to the examples just given, agencies did not list among their essential functions 20 of the 38 "high-impact" programs identified during the Y2K effort at the agencies we reviewed.

Another important consideration in identifying essential functions is the assessment of interdependencies among functions and organizations. As we have previously reported,[7] many agency functions rely on the availability of resources or functions controlled by another organization, including other agencies, state and local governments, and private entities. (For example, the Department of the Treasury's Financial Management Service receives and makes payments for most federal agencies.) The identification of such interdependencies continues to be essential to the related areas of information security and critical infrastructure protection. Although FPC 65 does not use the term "interdependencies," it directs agencies to "integrate supporting activities to ensure that essential functions can be performed."

Of the 34 plans we reviewed, 19 showed no evidence of an effort to identify interdependencies and link them to essential functions, which is a prerequisite to developing plans and procedures to support these functions and all other elements of COOP planning. Nine plans identified some key partners, but appeared to have excluded others: for instance, six agencies either make or collect payments, but did not mention the role of the Treasury Department in their COOP plans.

The high level of generality in FEMA's guidance on essential functions contributed to the inconsistencies in agencies' identification of these functions. In its initial guidance, FPC 65, FEMA provided minimal criteria for agencies to make these identifications, giving a brief definition only. According to FEMA officials, the agency is currently developing revised COOP guidance that will provide more specific direction on identifying essential functions. They expect the guidance to be released this Summer.

Further, although FEMA conducted several assessments of agency COOP planning between 1995 and 2001, none of these addressed the identification of essential functions. In addition, FEMA has begun development of a system to collect data from agencies on the readiness of their COOP plans, but FEMA officials told us that they will not use the system to validate the essential functions identified

[7] U.S. General Accounting Office, *Year 2000 Computing Challenge: Lessons Learned Can Be Applied to Other Management Challenges*, GAO/AIMD-00-290 (Washington, D.C.: Sept. 12, 2000).

by each agency or their interdependencies. According to these officials, the agencies are better able to make those determinations. However, especially in view of the wide variance in number and importance of functions identified, as well as omissions of high-impact programs, the lack of FEMA review lowers the level of assurance that the essential functions that have been identified are appropriate.

Additionally, in its oversight role, FEMA had the opportunity to help agencies refine their essential functions through an interagency COOP test or exercise. According to FPC 65, FEMA is responsible for coordinating such exercises. While it is developing a test and training program for COOP activities, it has not yet conducted an interagency exercise to test the feasibility of these planned activities. FEMA had planned a governmentwide exercise in 2002, but the exercise was cancelled after the September 11 attacks. It is currently preparing to conduct a governmentwide exercise in mid-May 2004.

Improper identification of essential functions can have a negative impact on the entire COOP plan, because other aspects of the COOP plan are designed around supporting these functions. If an agency fails to identify a function as essential, it will not make the necessary arrangements to perform that function. If it identifies too many functions as essential, it risks being unable to adequately address all of them. In either case, the agency increases the risk that it will not be able to perform its essential functions in an emergency.

Agency COOP Plans Addressed Some, but Not All, of FEMA's Guidance

As of October 1, 2002, almost 3 years after the planning deadline established by PDD 67, 3 of the agencies we reviewed had not developed and documented a COOP plan. The remaining 20 major federal civilian agencies had COOP plans in place, and the 15 components[8] that we reviewed also had plans.

[8] We reviewed 14 component plans: 1 plan covered a building that houses 2 components. Attachment III identifies the 15 components and the high-impact programs for which they are responsible.

However, after analyzing these plans, we found that none of them addressed all the guidance in FPC 65. Of the eight topic areas identified in FPC 65, these 34 COOP plans generally complied with the guidance in one area (developing plans and procedures); generally did not comply in one area (tests, training, and exercises); and showed mixed compliance in the other six areas. Specifically, when examining the governmentwide results of our analysis of the eight planning topics outlined in FPC 65, we found the following:

- *Essential functions.* Most agency plans identified at least one function as essential. However, less than half the COOP plans prioritized the functions, identified interdependencies among the functions, or identified the mission-critical systems and date needed to perform the functions.
- *Plans and procedures.* Most plans followed the guidance in this area, including a roster of COOP personnel, activation procedures, and the appropriate planning time frame (12 hours to 30 days).
- *Orders of succession.* All but a few agency plans identified an order of succession to the agency head. Fewer plans included orders of succession for other key officials or included officials outside of the local area in the succession to the agency head. Most plans did not include the orders of succession in the agency's vital records or document training for successors on their emergency duties.
- *Delegations of authority.* Few plans adequately documented the legal authority for officials to make policy decisions in an emergency.
- *Alternate facilities.* Most plans documented the acquisition of at least one alternate facility, and many include alternate facilities inside and outside of the local area. However, few plans documented that agencies had adequate space for staff, pre-positioned equipment, or appropriate communications capabilities at their alternate facilities.
- *Redundant emergency communications.* Most plans identified at least two independent media for voice communication. Few plans included adequate contact information or information on backup data links.
- *Vital records.* About one-quarter of plans fully identified the agency's vital records. Few plans documented the locations of all vital records or procedures for updating them.

- *Tests, training, and exercises.* While many agencies documented some training, very few agencies documented that they had conducted tests, training, and exercises at the recommended frequency.

Limitations in FEMA's Oversight Contribute to Noncompliance

The lack of compliance shown by many plans can be largely attributed to limited guidance and oversight of executive branch COOP planning. First, FEMA has issued little guidance to assist agencies in developing plans that address the goals of FPC 65. Following FPC 65, the agency issued more detailed guidance in April 2001 on two of FPC 65's eight topic areas: FPC 66 provides guidance on developing viable test, training, and exercise programs, and FPC 67 provides guidance for acquiring alternate facilities. However, it did not produce any detailed guidance on the other six topic areas.

In October 2003, FEMA began working with several members of the interagency COOP working group to revise FPC 65. Agency officials expect this revised guidance, which should incorporate the guidance from the previous FPCs and address more specifically what agencies need to do to comply with the guidance, to be released this summer. In addition, a member of the staff of the White House Homeland Security Council told us in March that the Council was also working on a new policy framework for federal COOP activities.

Second, as part of FEMA's oversight responsibilities, its Office of National Security Coordination is tasked with conducting comprehensive assessments of the federal executive branch COOP programs. With the assistance of contractors, the office has performed assessments, on an irregular schedule, of federal agencies' emergency planning capabilities:

- In 1995, it performed a survey of agency officials (this assessment predated FPC 65).
- In 1999, it assessed compliance with the elements of FPC 65 through a self-reported survey of agency COOP officials, supplemented by interviews.

- In 2001, it surveyed agency officials to ask, among other things, about actions that agencies took on and immediately after September 11, 2001.

Of these three assessments, only the 1999 assessment evaluated compliance with the elements of FPC 65. Following this assessment, FEMA gave agencies feedback on ways to improve their respective COOP plans, and it made general recommendations, not specific to individual agencies, that addressed programwide problems. However, it did not then follow up to determine whether individual agencies made improvements in response to its feedback and general recommendations. Besides inquiring about actions in response to the September 2001 attacks, the 2001 assessment was designed to provide an update on programwide problems that had been identified in the assessments of 1995 and 1999. FEMA did not address whether individual agency COOP plans had been revised to correct previously identified deficiencies, nor did it provide specific feedback to individual agencies.

According to FEMA officials, the system it is developing to collect agency-reported data on COOP plan readiness will improve its oversight. The system is based on a database of information provided by agencies for the purpose of determining if they are prepared to exercise their COOP plans, in part by assessing compliance with FPC 65. However, according to agency officials, while they recognize the need for some type of verification, they have not yet determined a method of verifying these data.

Without regular assessments of COOP plans that evaluate individual plans for adequacy, FEMA will not be able to provide information to help agencies improve their COOP plans. Further, if it does not verify the data provided by the agencies or follow up to determine whether agencies have improved their plans in response to such assessments, FEMA will have little assurance that agencies' emergency procedures are appropriate.

Agency officials attributed the limited level of oversight that we found to two factors. First, they stated that before its transition to the Department of Homeland Security, the agency did not have the legal or budgetary authority to conduct more active oversight of the COOP activities of other agencies. However, FPC 65 states that PDD

67 made the agency responsible for guidance, coordination, and oversight in this area, in addition to requiring agencies to develop COOP plans. Accordingly, although it cannot determine how agencies budget resources for such planning, it does have the authority to oversee this planning. Second, according to these officials, until last year, the agency devoted roughly 13 staff to COOP guidance, coordination, and oversight, as well as the development of FEMA's own plan. According to the official responsible for COOP oversight, the agency now has 42 positions authorized for such activities, 31 of which were filled as of December 31, 2003. The agency expects to fill another 4 positions in fiscal year 2004.

In summary, Mr. Chairman, while most of the agencies we reviewed had continuity of operations plans in place, those plans exhibited weaknesses in the form of widely varying determinations about what functions are essential and inconsistent compliance with guidance that defines a viable COOP capability. Agencies could experience difficulties in delivering key services to citizens in the aftermath of an emergency as a result of these weaknesses.

A significant factor contributing to this condition is FEMA's limited efforts to fulfill its responsibilities first by providing guidance to help agencies develop effective plans and then by assessing those plans. Further, FEMA has done little to help agencies identify those functions that are truly essential or to identify and plan for interdependencies among agency functions. FEMA has begun taking steps to improve its oversight, by developing more specific guidance and a system to track agency-provided COOP readiness information, and it is planning a governmentwide exercise. However, although the proposed guidance and exercise may help agencies improve their plans, the database that FEMA is developing to collect infromation on COOP readiness is weakened by a lack of planning to verify agency-submitted data, validate agency-identified essential functions, or identify interdependencies with other activities. Without this level of active oversight, continuity planning efforts will continue to fall short and increase the risk that the public will not be able to rely upon the continued delivery of essential government programs and services following an emergency.

In our report, we made several recommendations to the Secretary of Homeland Security to enhance the ability of the federal government to provide essential services during emergencies. In response to a draft of our report, the Under Secretary for Emergency Preparedness and Response agreed that better COOP planning is needed to ensure delivery of essential services, and that the department could do more to improve COOP planning. He added that FEMA has begun to correct the identified deficiencies and stated that the federal government is currently poised to provide services in an emergency. Once FEMA assesses and independently verifies the status of agencies' plans, it will have convincing evidence to support such statements about readiness in the future.

Mr. Chairman, this concludes my statement. I would be pleased to respond to any questions that you or other members of the Committee may have at this time.

Contacts and Acknowledgements

For information about this testimony, please contact Linda D. Koontz at (202) 512-6240 or at koontzl@gao.gov, or Mirko Dolak, Assistant Director, at (202) 512-6362 or dolakm@gao.gov. Other key contributors to this testimony include Barbara Collier, Neela Lakhmani, Susan Sato, James R. Sweetman, Jr., Jessie Thomas, and Marcia Washington.

Attachment I: Major Civilian Departments and Agencies Selected for Review

Department of Agriculture
Department of Commerce
Department of Education
Department of Energy
Department of Health and Human Services
Department of Housing and Urban Development
Department of Justice
Department of Labor
Department of State
Department of the Interior
Department of the Treasury
Department of Transportation
Department of Veterans Affairs
Agency for International Development
Environmental Protection Agency
Federal Emergency Management Agency
General Services Administration
National Aeronautics and Space Administration
National Science Foundation
Nuclear Regulatory Commission
Office of Personnel Management
Small Business Administration
Social Security Administration

Attachment II: 38 High-Impact Programs and Responsible Agencies Included in Our Review

Agency	High-impact programs
Department of Agriculture	Food safety inspection
	Child nutrition programs
	Food stamps
	Special supplemental nutrition program for women, infants, and children
Department of Commerce	Patent and trademark processing
	Weather service
Department of Education	Student aid
Department of Energy	Federal electric power generation and delivery
Department of Health and Human Services	Disease monitoring and warnings
	Indian health services
	Medicaid
	Medicare
	Organ transplants
	Child care
	Child support enforcement
	Child welfare
	Low income home energy assistance
	Temporary assistance for needy families
Department of Housing and Urban Development	Community development block grants
	Housing loans
	Mortgage insurance
	Section 8 rental assistance
	Public housing
Department of Justice	Federal prisons
	Immigration
Department of Labor	Unemployment insurance
Department of State	Passport applications and processing
Department of the Interior	Bureau of Indian Affairs programs
Department of the Treasury	Cross-border inspection services
Department of Transportation	Air traffic control system
	Maritime search and rescue
Department of Veterans Affairs	Veterans' benefits
	Veterans' health care
Federal Emergency Management Agency	Disaster relief
Office of Personnel Management	Federal employee health benefits

Agency	High-impact programs
	Federal employee life insurance
	Federal employee retirement benefits
Social Security Administration	Social Security benefits

Source: GAO analysis of OMB guidance.

Attachment III: Component Agencies Reviewed, with High-Impact Program Responsibilities

Department	Component	High-impact programs
Department of Commerce	National Oceanic and Atmospheric Administration	Weather service
	Patent and Trademark Office	Patent and trademark processing
Department of Health and Human Services	Centers for Disease Control and Prevention	Disease monitoring and warnings
	Center for Medicare and Medicaid Services	Medicare and Medicaid
	Food and Drug Administration	Organ transplants
	Indian Health Service	Indian health services
Department of Housing and Urban Development	Government National Mortgage Association	Housing loans
	Office of Community Planning and Development	Community development block grants
	Office of Housing	Section 8 rental assistance and mortgage insurance
	Office of Public and Indian Housing	Public housing
Department of the Interior	Bureau of Indian Affairs	Indian affairs programs
Department of the Treasury	U.S. Customs Service	Cross-border inspection services
Department of Transportation	Federal Aviation Administration	Air traffic control system
	U.S. Coast Guard	Maritime search and rescue
Department of Veterans Affairs	Veterans Benefits Administration	Veterans' benefits

Source: GAO analysis of OMB guidance.

Chairman TOM DAVIS. Linda, let me just start. The bottom line is, are agencies really prepared for the worst?

Ms. KOONTZ. Agencies do not have plans at this point that are fully compliant with the requirements of FPC 65 and therefore I'd have to conclude that there is no assurance that they are prepared for an emergency.

Chairman TOM DAVIS. In fact, some of them are fairly woefully prepared.

Ms. KOONTZ. That is correct.

Chairman TOM DAVIS. You report that 19 agencies failed to identify their interdependence with other agencies and how these interdependencies affect their essential functions. Was GAO provided with an explanation as to why these agencies didn't identify their interdependency in COOP plans?

Ms. KOONTZ. I don't—excuse me. I think part of the issue that my staff is telling me that the requirement to identify interdependencies, we think, would be a good practice. But that requirement is not specifically outlined in FPC 65. So that is most likely the reason.

Chairman TOM DAVIS. All right. Do you get the feeling some of these agencies are just checking the box? This is just another requirement that they have to do? This isn't really—this isn't part of their mission, but it's paperwork they have to turn in so it's kind of—they're not utilizing the resources; they're putting them toward other missions in the department?

Ms. KOONTZ. It's hard for me to comment on a specific agency's motivation for what they do. But we have to say that in some cases we saw what we thought looked like sort of a rote or a template approach to the development of plans.

Chairman TOM DAVIS. Yes. I think one of the difficulties is, both from the executive branch and from the legislative branch, we put all of these different requirements on agencies, and it's hard for them to sort out what their priorities are. If they do them all, they'd never be able to get anything done. And so as a result of that, sometimes nothing gets done.

One of the rules of this committee is to kind of highlight shortcomings in some of these areas. This area, cybersecurity area, again, another one similar, where agencies check boxes but don't really make this mission-critical. And they may be able to escape with this. This is one of those issues that, you know, hopefully we will never see that kind of disaster and it will never happen. But if it does, and we are not prepared, of course the results then are worse by an exponential amount.

Ms. KOONTZ. And if I could add to that, the fact that FEMA hasn't done the regular checking and oversight of the plans, I think that created part of the situation that you see today. If agencies realize that someone's going to be routinely looking at these plans, I think that would provide greater incentive for providing resources for this activity.

Chairman TOM DAVIS. And there is no requirement, is there, that they send the plans to Congress? They send them up through FEMA, right?

Ms. KOONTZ. No, sir.

Chairman TOM DAVIS. That might be something we can get access to, that we look at to try to underscore the importance of this. I mean, hopefully again, this is something if you don't do it, it'll never happen, nobody will know the difference. But if you have a disaster, there we are.

The report states that FEMA attributed its lack of oversight of these plans in part to its limited number of personnel responsible for guidance. Now, as a result of your investigation, can GAO concur with FEMA the inadequate personnel numbers significantly affected FEMA's ability to conduct oversight?

Ms. KOONTZ. We didn't specifically evaluate the numbers of staff that would be necessary for FEMA to conduct this oversight activity. However, we do know that FEMA has, since we completed our work, undertaken a rather large effort to get many more people involved. So this should not be a problem going forward.

Chairman TOM DAVIS. OK. Thank you very much.

Ms. Watson.

Ms. WATSON. The Chair asked a question about if there was a requirement to report to us, and I'd like you to describe what you think we should know in advance so that as we go about budgeting for whatever, there could be appropriate resources there to address what might occur. We really need to start looking ahead. We've had the shock of an experience that we will never forget now. How do we—we're new at this, and I understand that. We were caught in a blind spot. Unready. But what is it going to mean in terms of resources to be ready? Do you have a comment?

Ms. KOONTZ. A couple of parts to that question. I think in terms of resources that, according to the report we saw from OMB on combating terrorism that was published in September 2003, apparently it's not unusual for agencies to spend several million dollars working on continuity of operations planning. And indeed the President asked for over $100 million for this purpose in 2004. I would have to followup to tell you what was actually devoted, however.

In terms of reporting to Congress, I think that one of the things that Mr. Davis has asked us to do is to set a baseline of continuity of operations planning efforts, which we have done with our first report. And in following up on that, hopefully you'll be able to see the changes that take place over time and to be able to influence those changes further.

Ms. WATSON. Thank you.

Chairman TOM DAVIS. Thank you very much. Gentlelady from Tennessee.

Mrs. BLACKBURN. Thank you, Mr. Chairman.

And thank you for taking the time to be here and visit with us today. You mention in your report the Y2K efforts, and my assumption—which I would like to know if it's correct or not—is that where you have drawn your baseline, as working from the efforts that were made there in preparation for Y2K, that helps with your baseline?

Ms. KOONTZ. What we drew from the Y2K effort was the previously identified list of 38 essential functions that were identified specifically for that purpose. And we use this as an example against which to evaluate plans to see if these essential functions were present or not. We don't mean to imply that this is the definitive list of essential functions, but we felt it was one strong example of where the government had already identified programs that had a high impact on the public.

Mrs. BLACKBURN. OK. Now, have you required the different agencies and departments to—going into those and looking at that Y2K planning and into those agencies and programs, have you required them to go on and give you the coordination with State and local agencies for implementation of continuing services as it affects those departments?

Ms. KOONTZ. We haven't yet looked at the issue of coordination between the Federal and the State and local governments.

Mrs. BLACKBURN. OK. What is the status of the agency's information technology that is needed to oversee these essential functions?

Ms. KOONTZ. Well, one of the aspects of any kind of continuity planning would be to assure that your critical infrastructure and your systems would be available in an emergency, and this would also extend to what we call vital records as well. In order to operate in an emergency situation, one has to have access to the information that is needed for decisionmaking. So these are all aspects of continuity of operations planning. What we saw among the agencies was, frankly, mixed preparedness in all these areas.

Mrs. BLACKBURN. Thank you.

Chairman TOM DAVIS. Thank you very much.

Ms. Norton.

Ms. NORTON. Thank you very much, Mr. Chairman. Thank you for calling this hearing. It is an interesting hearing coming up today, especially since we have on the floor a continuity of operations bill.

I'm confused even by the GAO report, because all the dates I see goes back to 1999. You speak on the first page about assessments of agency compliance conducted in 1999 and none conducted since then. And again, at page 5, a reference to July 1999, and the assessments conducted to address any emergency or situation that could disrupt normal operations, including localized emergencies.

Well, I'm really wondering whether anything that goes back to 1999 is relevant at all. That is to say, with the intervention of September 11th, I'm not sure what FEMA would be reviewing, if FEMA is reviewing plans that were set in motion in 1999, when on page 5 of your own report you say it relates to any emergency, including localized emergencies. I just wonder whether they don't need to start all over again, whether any plan that was prepared before September 11th is worth the paper it's written on, whether or not we don't need fresh eyes when we look at what a local emergency is when we look at infrastructure. So I would like some sense from you whether you think we can actually pick up from 1999 or whether we ought not step back and essentially begin again.

Ms. KOONTZ. I can clarify a little bit. The requirements first came into being in 1999, and agencies were required to have a continuity of operations plan in place at this point for that same year. It was also the same year that FEMA did an assessment of plans and gave agencies feedback as to strengths and weaknesses.

Ms. NORTON. I think Oklahoma may have occurred by that time, so I'm sure there was some sense that you could get, you know, a large emergency. But go ahead.

Ms. KOONTZ. So that was the first round of plans. But I wouldn't want to lead you to believe that none of those plans have been updated since 1999. Some agencies have taken steps to review their plans once or twice since then, but it varies quite a bit across the board. Certainly anything that went back to 1999 would need a significant reassessment before it could be brought up to date, and indeed we found that regardless of when the plan had been prepared, that most of them did not hit the majority of the requirements. In fact, we found not a single one that met all the requirements in their entirety. So all of them need a significant relook at this point. But I just wouldn't want you to believe that nothing has happened since then.

Ms. NORTON. Well, obviously, at the agency level one would need to particularize what the emergency planning was. I have no confidence that you begin by saying, hey, agencies, figure out what to do. I don't understand why there shouldn't be some overall—you talk in your report about the great disparities among these agencies. Much of that is to be expected. But without FEMA's guidance as to what constitutes a plan, what else could you expect? So I don't see how we can go back and criticize the agencies or even criticize FEMA for not going back agency by agency.

My question is, why isn't there some general guidance as to what minimally an agency should be doing, its plan should be, with the agencies filling in the particulars, rather than this kind of ground-up approach and then us criticizing the agencies? Because somehow they are very different from agency to agency, as if that isn't exactly what you should expect if you haven't given agencies some idea of what continuity of operation should be all about.

So, Mr. Chairman, I must say that I appreciate your calling this hearing, but I think we are just going at it the entirely wrong way to say to agencies out there, hey, you all come up with what you should be doing to continue operations. Without some general guidance as to "these are the basics, now fill in" does not give me confidence, particularly here in the National Capital Region, that if there were an emergency it could be handled.

Chairman TOM DAVIS. Well, doesn't FPC 65 give out the basic guidance?

Ms. KOONTZ. Yes. FPC 65 provides basic guidance on the eight elements of a viable COOP capability.

Ms. NORTON. And isn't that also from 1999?

Ms. KOONTZ. Yes, that is from 1999.

Ms. NORTON. Well, that is my problem. I think the world has changed since September 11, 2001, and that was before 1999. That was after 1999.

Chairman TOM DAVIS. Ms. Koontz, any response?

Ms. NORTON. I think that is a more radical critique than the GAO report is what I'm trying to say.

Chairman TOM DAVIS. OK. Ms. Koontz.

Ms. KOONTZ. I would just say that one of the things that we point out I think quite strongly in our report is that the identification of essential functions is a very critical first step in doing effective continuity planning. If you don't do that right, it probably doesn't matter what do you after that because you haven't figured out what it is you need to deliver in an emergency.

But we also point out that the guidance to agencies, although they have issued general guidance, it was not specific enough to agencies for them to identify really what an essential function was and get any consistency across agencies; and that was compounded by the fact that FEMA was not doing the regular kind of checking and oversight to provide their expertise, to lend their expertise to the development of these plans and provide their broad view of what was going on government-wide. So I think our report does address some of the issues that you're identifying here.

Ms. NORTON. Well, thank you very much; and thank you Mr. Chairman.

Chairman TOM DAVIS. Thank you very much.

Mrs. Davis.

Mrs. JO ANN DAVIS OF VIRGINIA. Thank you, Mr. Chairman; and thank you, Ms. Koontz, for being here.

You know, it seems to me that if we had the technology in place for telecommuting that in the event of an attack here in Washington, for instance, people could work at home. So I guess my question is, have any of the agencies—when you reviewed their plans, had they considered or included telecommuting in their continuity of operations plans? Because that's been the hardest thing. We've been—I mean, we have tried to get agencies to allow telecommuting, and it seems as hard as pulling teeth sometimes.

Ms. KOONTZ. Uh-huh. And using both the use of alternatives facilities and the use of telecommuting could be a reasonable strategy to use in continuity of operations planning, depending on the kind of emergencies that we're talking about.

Mrs. JO ANN DAVIS OF VIRGINIA. Well, if we had the agency allowing the telecommuting now, it would be in place; and then there would be an answer to some of the problems for some of these agencies.

My other question—you know, I heard you say that if FEMA or someone were doing reviews or what have you, then these agencies might get off the stick, I guess, is what you meant. And it bothers me a little bit, because are you saying then that our agencies don't do what we tell them to do unless they know we are going to check on them?

But my real question to you—I mean, that was just a side note. It bothers me to hear that. But did agency personnel responsible for developing the continuity of operation plans indicate why they have not followed the guidelines that FEMA gave them? I mean, the person in each agency who is responsible, did they give you any feedback?

Ms. KOONTZ. Well, there are a couple of different classes of things going on here.

I think, first, in some cases the guidance isn't very clear; and so agencies maybe tried to implement it the best they could, but it was predictably then inconsistent across the government. So you have some of that going on.

In other cases, I think agencies told us that they had prepared their plan. It had been reviewed by FEMA in 1999. They thought the feedback they received was that plan was all right; and, frankly, I think they were surprised in some instances when we said, well, we don't think this meets the requirements or the guidance of FPC 65.

So there was a couple of different kinds of things going on there.

Mrs. JO ANN DAVIS OF VIRGINIA. Sounds like a communication problem, Mr. Chairman. It seems like we have that a lot in the Federal Government. I don't know how we can fix that.

But thank you so much, and I would strongly suggest that we push the telecommuting if we can.

Chairman TOM DAVIS. Great. I think Mrs. Davis' idea on the telecommuting is something that for agencies here we need to do more of. I mean, this committee will hold followup hearings on that. Obviously, if an office gets devastated, people don't need to be in the office in many cases to carry out their duties.

Mr. Van Hollen.

Mr. VAN HOLLEN. Nothing.

Chairman TOM DAVIS. No questions.

Mrs. Maloney.

No questions.

Thank you very much. This has been very helpful for us. We may have some followup pending some of the others, but we appreciate your oversight on this and your analysis.

Ms. KOONTZ. Thank you.

Chairman TOM DAVIS. Thank you very much.

We will proceed now to our second panel. I am going to thank Under Secretary Michael Brown, the Honorable Michael Brown, the Under Secretary for Emergency Preparedness and Response Directorate from the U.S. Department of Homeland Security, for being with us today. Why don't we take a minute recess, but I'll wait for him to come in.

There he is. Mr. Secretary, thank you for being with us. Why don't you stay—and I'll swear you in, our policy.

[Witness sworn.]

Chairman TOM DAVIS. Thank you very much for being with us today.

We'll have some lights in front of you, the panel. After 4 minutes, an orange light will come up, giving you a minute to make it 5. If you feel you need to go over it, we're not pressed for time. We'll do that. But your entire testimony is part of the record, and our questions have been based on that.

But thank you very much for being with us today, and thank you for the job you're doing.

STATEMENT OF MICHAEL BROWN, UNDER SECRETARY FOR EMERGENCY PREPAREDNESS AND RESPONSE DIRECTORATE, U.S. DEPARTMENT OF HOMELAND SECURITY

Mr. BROWN. Thank you, Mr. Chairman.

Chairman TOM DAVIS. You have to make sure the mic is on. That's the toughest part of the whole thing.

Mr. BROWN. I'm not used to coming in second. I guess you're just ready. Go ahead and start then, right? OK.

Good morning, Chairman Davis and members of the committee. My name is Michael D. Brown, and I am the Under Secretary for Emergency Preparedness and Response.

Chairman TOM DAVIS. Mr. Brown the reason we have you second is we have GAO first and we give you the last word.

Mr. BROWN. Sure. Right.

Chairman TOM DAVIS. So it's really to your advantage to be in that position.

Mr. BROWN. Great.

Thank you for the opportunity to appear before you today to discuss the Federal Emergency Management Agency's role in supporting the Nation's Continuity of Operations and its program.

FEMA was designated the lead agency for Continuity of Operations for the Federal executive branch by Presidential guidance on October 21, 1998. Among other things, this guidance requires Federal agencies to develop Continuity of Operations plans to support their essential functions. FEMA's leadership role is to provide guidance and assistance to the other Federal departments and agencies in this important area. We have taken this responsibility very seriously and have worked hard to provide this guidance.

As the program expert for the Federal executive branch COOP activities, FEMA and the Department of Homeland Security have made significant strides toward ensuring that COOP plans exist at all levels of departments and agencies. This effort entails our involvement with hundreds, if not thousands, of various COOP plans and close coordination with the General Services Administration.

We have aggressively developed working relationships across the government—to include the legislative and judicial branches—to expend our efforts at providing advice and assistance to other Federal departments and agencies in the COOP arena. We have established numerous interagency COOP working groups at the headquarters and at the regional levels. These working groups have opened communication channels across the government regarding COOP plans and programs and have helped organizations develop more detailed COOP planning in order to leverage capabilities and to improve interoperability. Moreover, we have developed new COOP testing, training and exercise programs to help ensure that all departments and agencies are prepared to implement their COOP plans.

Significantly in fact—FEMA tested its own COOP plan and capabilities in December 2003 by conducting Exercise Quiet Strength. This headquarters COOP activation involved the notification and relocation of nearly 300 FEMA personnel on our emergency relocation group, and it successfully demonstrated our ability to perform FEMA's essential functions from an alternate site under emergency conditions.

We are now leading the interagency Exercise Forward Challenge scheduled for next month. This full-scale COOP exercise will require departments and agencies in the National Capital Region to relocate and operate from their alternate facilities. Some 45 departments and agencies plan to participate in Forward Challenge. A prerequisite for their participation is for each department and agency to develop their own internal Forward Challenge COOP exercise. As a result, there will be approximately 45 separate but linked COOP exercises conducted concurrently with the main Forward Challenge event. Because of these internal exercises, Forward Challenge preparation has cascaded across the country, with departments and agencies as far away as Fort Worth and Seattle participating.

Our support for COOP exercises and training is not limited to the Washington, DC, area. Working with the Federal executive boards, FEMA has conducted interagency COOP exercises in Denver and Chicago; and additional exercises are scheduled in Kansas City on April 29 and in Houston on June 14. To help facilitate this effort, FEMA has developed a generic interagency COOP exercise template that can be easily adapted for use in the field.

Mr. Chairman, you have specifically asked me to address what steps FEMA is taking to address each of the executive agencies' COOP plans and what steps we are taking to address deficiencies in those plans. Through our strong working relationships and through new and ongoing COOP initiatives, we are leading the government's COOP program to ensure improved coordination and provide enhanced planning guidance. FEMA established the Interagency COOP Working Group in the National Capital Region comprised of 67 separate departments and agencies. This working group includes the Library of Congress, the GAO, U.S. Senate, the D.C. Department of Transportation, the U.S. court systems and the Metropolitan Washington Council of Governments. At the regional level, FEMA has used a phased approach to establish COOP working groups with many of the Federal executive boards and Federal executive associations across the country.

In addition, we are revising the Federal preparedness circular for COOP. The goal is to have a single-source document that all departments and agencies can refer to for their COOP programs. The new Federal preparedness circular incorporates many of the GAO's recent recommendations for improvements. It includes detailed information on how to identify essential functions and discusses the importance of interdependencies between departments and agencies.

Mr. Chairman, the ability of the Federal Government to deliver essential government services in an emergency is of critical importance. In June, we agreed that improved planning was needed to ensure the delivery of essential services. However, I unwaveringly believe the Federal Government is currently poised to deliver those services in an emergency that requires the activation of COOP plans.

Mr. Chairman, thank you for you time; and I'll be happy to answer any questions you may have.

Chairman TOM DAVIS. Thank you very much.

[The prepared statement of Mr. Brown follows:]

Statement of Michael D. Brown
Under Secretary
Emergency Preparedness and Response Directorate
Federal Emergency Management Agency
Department of Homeland Security

Before the
Committee on Government Reform
House of Representatives
on
FEMA's Continuity of Operations Program Management

April 22, 2004

Good morning, Chairman Davis and members of the Committee. I am Michael D. Brown, the Under Secretary for Emergency Preparedness and Response of the Department of Homeland Security. Thank you for the opportunity to appear before you today to discuss the Federal Emergency Management Agency's role and activities to support the Nation's Continuity of Operations (COOP) planning and program.

The Federal Emergency Management Agency (FEMA) was designated as Lead Agent for Continuity of Operations (COOP) planning for the Federal Executive Branch by Presidential guidance on October 21, 1998. Among other things, the Presidential guidance requires Federal agencies to develop Continuity of Operations Plans to support their essential functions. FEMA's role is to provide guidance and assistance to the other Federal Departments and Agencies in this important area. We have taken this responsibility very seriously and have worked hard to provide this guidance through development of Federal Preparedness Circulars (FPCs), establishment of interagency COOP working groups, and other supporting activities. Today I would like to highlight FEMA's role in leading the Federal government's ability to ensure the delivery of essential government services in an emergency through improved COOP planning coordination, enhanced training opportunities, and robust exercise and assessment programs.

As the program expert for the Federal Executive Branch COOP activities, FEMA and the Department of Homeland Security have made significant strides toward ensuring that COOP plans exist at all levels of Departments and Agencies, at the Washington, D.C. headquarters level, and at Regional and Field offices across the country. This effort entails our involvement with hundreds, if not thousands, of various COOP plans and close coordination with the General Services Administration to effectively execute our COOP roles and responsibilities. We have aggressively developed working relationships across the government – to include the Legislative and Judicial Branches – to expand our efforts at providing advice and assistance to other Federal Departments and Agencies in the COOP arena; helping them comply with guidance in Federal Preparedness Circular – 65, the Federal Executive Branch's COOP implementation guidance document; and establishing numerous interagency COOP working groups at the headquarters and regional levels. These working groups have opened communication channels across the government regarding COOP plans and programs, and have helped organizations develop more detailed COOP planning in order to leverage capabilities and improve interoperability. Moreover, we have developed new COOP test, training, and exercise programs to help ensure that all departments and agencies are prepared to implement their COOP plans.

Significantly, and for the first time, FEMA tested its own COOP plan and capabilities in December 2003 by conducting Exercise Quiet Strength. This headquarters COOP activation involved the notification and relocation of nearly 300 FEMA personnel on our Emergency Relocation Group and successfully demonstrated our ability to perform FEMA's essential functions from an alternate site under emergency conditions. Exercise Quiet Strength is the first of what will be, at least, a biennial exercise of FEMA's COOP plan.

As you are no doubt aware, our Office of National Security Coordination is leading Exercise Forward Challenge scheduled for next month. This full-scale COOP exercise will require departments and agencies in the National Capital Region to relocate and operate from their alternate facilities. We began preparing for this exercise in May 2003, and have conducted numerous COOP related planning, training, and preparedness activities. As a result of our nearly year - long efforts, 38 department and agency headquarters have FEMA trained COOP exercise design teams contributing to Forward Challenge. Also, in preparation for this exercise, FEMA established and tested an Executive Branch COOP alert and notification system that is capable of contacting departments and agencies 24 hours a day, 7 days a week. Here, I would like to point out some forty-five departments and agencies plan to participate in Forward Challenge. A prerequisite for their participation in the exercise is for each department and agency to develop their own internal Forward Challenge COOP exercise. As a result, there will be approximately forty-five separate, but linked, COOP exercises conducted concurrently with the main Forward Challenge event. Because of these internal exercises, Forward Challenge preparation has cascaded across the country with departments and agencies as far away as Fort Worth, Texas and Seattle, Washington participating. In addition, a communications test will be conducted as part of Forward Challenge to test communications interoperability to, and between, participating agencies' alternate relocation sites. Approximately, fifty-two separate communications checks for each participating organization are scheduled during the exercise, in addition to the interagency communication and coordination required by the Forward Challenge Master Scenario Events List. This government – wide exercise, like FEMA's own Quiet Strength exercise, will become a biennial training event that will provide a means to assess Department and Agency COOP capability.

Our support for COOP exercises and training is not limited to the Washington, D.C. area. Working with the Federal Executive Boards (FEBs), FEMA has conducted interagency COOP exercises in Denver and Chicago, and additional exercises are scheduled in Kansas City on April 29th and Houston on June 14th. To help facilitate this effort, FEMA has developed a generic interagency COOP exercise template that can be easily adapted for use in the field. These exercises are an important component of the COOP preparedness and assessment process, as they assist Federal departments and agencies with the identification of strengths and weaknesses in their COOP plans and programs. We believe this COOP exercise program provides departments and agencies across the country a format and methodology to ensure plans are completed, tested, and assessed to determine compliance with the standards outlined in Federal Preparedness Circular - 65. More importantly, it helps ensure the Federal government's ability to deliver essential services during any emergency.

Mr. Chairman, you specifically asked me to address what steps FEMA is taking to assess each of the Executive agencies' COOP plans and what steps we are taking to address deficiencies in those plans. Through our strong working relationships and through new and ongoing COOP initiatives, we are leading the government's COOP program to ensure improved coordination and provide enhanced planning guidance. In that regard, FEMA established the Interagency COOP Working Group in the National Capital Region. It is comprised of 67 separate departments and agencies. Also participating in this Interagency COOP Working Group are COOP planners from the Library of Congress, the GAO, the U.S. Senate, the D.C. Department of Transportation, the U.S. Courts, and the Metropolitan Washington Council of Governments. This Interagency Working Group meets monthly and provides an excellent forum for sharing ideas on the spectrum of COOP preparedness issues. FEMA has also encouraged departments and agencies to establish their own internal COOP Working Groups. At the Regional level, FEMA has used a phased approach to establish COOP working groups with many of the Federal Executive Boards and Federal Executive Associations across the country. FEBs in New York City, Boston, Chicago, Philadelphia, Oakland/San Francisco, Houston, Albuquerque, Kansas City, Denver, Atlanta, and Seattle are examples of where FEMA has begun the process of establishing COOP Working Groups as forums for providing COOP program advice and guidance. In addition, we are revising the Federal Preparedness Circular for COOP, combining all previous COOP FPCs and including new and more definitive planning guidance into this circular. The goal is to have a single-source document that all departments and agencies can refer to for their COOP programs. The new FPC incorporates many of the GAO's recent recommendations for improvement and includes detailed information on how to identify essential functions, and discusses the importance of interdependencies between departments and agencies. To further assist the departments and agencies in their COOP planning, the new FPC also provides examples of high impact programs that should be considered agency essential functions and stresses the importance of prioritizing these functions. It includes detailed guidance on how agencies should develop their Orders of Succession and Delegations of Authority, as well as information on identification and preparation of alternate facilities. Guidance on establishing redundant emergency communications, identifying vital records and databases, and development of COOP test, training, and exercise activities is also included. This revised document includes and updates all previous guidance that had been included in FPC 66, Test, Training, and Exercise Program for Continuity of Operations, and FPC 67, Acquisition of Alternate Facilities for Continuity of Operations. It also includes new, detailed information on six additional COOP elements outlined in FPC 65 and mentioned in the recent GAO report on continuity of operations.

I would also briefly mention the COOP Readiness Reporting System that we have begun fielding. This classified system will assist us in providing more accurate and timely information on government – wide COOP capabilities. Should you find it necessary, we can provide you with a classified briefing on this program.

Mr. Chairman, the ability of the Federal government to deliver essential government services in an emergency is of critical importance. In general, we agree that improved planning is needed to ensure delivery of essential services. However, I believe the Federal government is currently poised to provide those services in an emergency that requires the activation of COOP plans. This capability was effectively demonstrated during Hurricane Isabel and during the holidays' elevated Code Orange alert. In both instances we were very pleased to find that all Departments and agencies were well postured to immediately transition operations to alternate facilities.

Thank you for your time and I will be pleased to answer any questions you may have.

Chairman TOM DAVIS. Just to start off, I was pronouncing it COOP plans. You're pronouncing it the COOP plans. And the reason I called it COOP is because chickens are in charge of the COOP, and I didn't want anyone in the administration to cry foul of what I was doing, which is eggsactly what they do. I mean, obviously, we don't want any agency——

Mr. BROWN. I can't compete with this humor.

Chairman TOM DAVIS [continuing]. We don't want the agencies winging it on their COOP plans. So we will risk ruffling some feathers here today. But I think it's fair to say the administration's proposal so far are nothing to crow about.

But let me ask a few questions.

Mr. BROWN. OK. Because I'm ready to fly the coop, so——

Chairman TOM DAVIS. Everybody acknowledges that the first and most critical element of any COOP planning is the identification of every essential function that an agency performs and will attempt to maintain in case of an emergency. But GAO reports that individual agencies' identification of essential functions really vary widely. Can you just kind of review in brief for us what steps FEMA has taken to assure that these critical functions are accurately carried out by every Federal agency?

Mr. BROWN. Absolutely, Mr. Chairman. FEMA has a coordination role and provides guidance and assistance, but it is really up to the departments and the agencies themselves to determine what's essential for their COOP plans.

We do such things as having a monthly forum through the Interagency COOP Working Group for departments and agencies to address those issues and insure best practices.

I also believe that the revised preparedness circular that is soon to be released at the end of the fourth quarter will provide better decisionmaking guidance to the departments and agencies that will also ensure consistency across the Federal Government.

Moreover, through a readiness reporting system that FEMA is now implementing, we will be in a better position to provide more accurate and timely information regarding each department and agency's COOP activities.

But I believe it's important to note—particularly important to note that, for the first time ever, as I said in my oral statement, FEMA exercised its headquarters COOP plan. It involved the actual notification and actual deployment of our emergency relocation group to our alternate facility. This is the first time ever that FEMA has done that. And that we will now oversee for the first time ever a Federal Government-wide COOP exercise that will allow us to establish a baseline for future exercises that we want to have now on an annual basis.

Chairman TOM DAVIS. I asked this question of the previous panel. Are agencies prepared for the worst today? Or are we getting there?

Mr. BROWN. We are certainly getting there. And my hesitation is not about preparedness. My hesitation, Mr. Chairman, is about what is the worst—because the worst, in my world, unfortunately, is, you know, the detonation, for example, of a nuclear device or a dirty bomb or a bioterrorist event which will result in catastrophic casualties and a catastrophic disaster of proportions that will overwhelm all of us. So that is the reason for my hesitation. I believe that every department and agency has a very good, robust COOP plan in place that we just now need to fine-tune.

Chairman TOM DAVIS. I mean, the experience of this committee as we go through little emergencies that come across the city—for example, recently we had tractor man. We had a guy on a tractor hold up traffic and tie up this city for three rush hours. And there was—the planning that took—there was no planning. There was a division over really what the priorities were to make sure that the person escaped—I mean, that he wasn't injured and was apprehended, that no one was injured. Nobody looked out for—and so some of this stuff gets very contradictory as you start to have to go down the path and decide what the priorities. You can't anticipate any and all bad things that can happen.

Mr. BROWN. No, but I think, based on the template that we have put together or the revision of the Federal preparedness circular, that we will be able to provide them with a template that allows them to respond to any—almost any kind of disaster.

Chairman TOM DAVIS. Ms. Norton's concern in the previous panel was that we were dealing with a circular from the executive that came—was a 1999 circular, before September 11. On September 11 I can tell you we certainly weren't prepared on Capitol Hill. I mean, we didn't know who to call. We were kind of irrelevant to the process, though, basically. I mean, we don't like to think of this that way, but the government went on fine. Everybody—the military did their job. The police did their job. Other agencies kicked in. It is a lot more important than what happens here.

I guess my question is, as we look at different agencies we see different levels of planning for this. That's not unusual. What you usually find is we put so many requirements on these different agencies and secretariats and the like that they have to sort it out in some, take it more seriously than others. In fact, some of them, how they plan is going to be more important to the American people than others.

So as you look over this in terms of your planning and the checklists and everything else, what are we doing to check on this?

There was an allegation at GAO that maybe you didn't have enough people to really implement this job. This is a contingency planning, so it may never happen, and some agency leaders, I think, think, well, I don't have to do this because it'll never happen, and then I can put my resources somewhere else and accomplish something that everybody—that I know will happen. What's your reaction to that?

Mr. BROWN. Let me—three things I want to respond to, Mr. Chairman.

First of all, your comment about the ability of the Federal Government—the ability of executive branch to be able to actually COOP and respond in terms of an emergency. The good news I believe out of this hearing should be that all of the major departments and agencies—in fact, all the departments and agencies have a COOP plan in place that we have reviewed and we have looked at.

Do these need to be fine-tuned? Absolutely. Do we need to continue to improve those? Absolutely. But there is no place in the executive branch of the departments and agencies where there is a lack of a COOP plan. So that's the good news.

GAO is correct in that we have been concerned about the staffing levels. But one of the priorities that I have put in since I have become the Under Secretary was to increase the staffing in our national security office, coordination office and we have increased the staff levels. Additionally, we have received incredible support from President Bush and the administration and in the 2005 budget there is a $12 million increase, specifically for COOP activities.

Chairman TOM DAVIS. The other criticisms—one other criticism that came out of the GAO report—I wouldn't call it criticism, but one of their observations was that some of the COOP reports that came in really didn't talk about how they interact with other agencies, that they simply look at what they did. And it was almost like a checklist, which, by the way, is not uncommon. I'm not trying to be overly critical here. I'm just trying to make sure that as we look forward we can continue to improve.

Mr. BROWN. And that is exactly one of the things that we want to test in Exercise Forward Challenge. It's not just their ability to pick up and move and go to their alternate sites, but how do they interact, how are the interdependencies, how is the interoperability of communications among the different Departments and Agencies and where can we improve on that. So you have identified exactly one of the areas that we intend to push in the exercise.

I would just take this opportunity also to caution everyone about the exercise, because it is my philosophy, and it is one that I'm trying to push all the way through FEMA and the entire departments, that we don't do exercises to make things look good. We do exercises to push the envelope, to find out where the vulnerabilities are, to find out where the weaknesses are so that we can come back and improve upon them. So I fully expect after Exercise Forward Challenge for us, the Interagency Working Group to get back together and find places where interdependencies didn't exist, and we need to improve those. That's the purpose of the exercise.

Chairman TOM DAVIS. And you did provide information about Exercise Quiet Strength, which was FEMA's December 2003, exercise to test its headquarters COOP plan. But that is an isolated exercise of one agency; and in reality, of course, particularly with you all, an actual emergency would involve government-wide functions. Is there an effort to test some of that later on in the interaction of some of the agencies?

Mr. BROWN. There absolutely is. But before we can go out and be a good leader and convince all the other departments and agencies to do this we have to show that we are willing to do it, too. And since FEMA had never done this exercise I was very pleased that we were able to pull it off and be as successful as we were.

Chairman TOM DAVIS. Thank you very much.

Ms. Watson.

Ms. WATSON. Thank you for being here and helping us get our—wrap our minds around how comprehensive this emergency preparedness and the planning might be.

I was just given a printout from the L.A. Times that this is the third loss of power at the Los Angeles Airport in 10 days. When you think about Los Angeles Airport under the FAA being one of the major ports on the west coast, it's very troubling to know that a bird can stand on a wire, spread its wings, make the connection and out the whole airport and all the flights all the way, nationally and internationally—the third time in 10 days.

My question is, in your interagency efforts, is it the FAA, then, that would be able to take a look at all of our airports? It seems to me that, you know, I can't really understand how a bird could do this three times in 10 days and where our backup systems are.

Mr. BROWN. I assume it wasn't the same bird.

Ms. WATSON. No. I think that bird has been—is toast. But, you know, it just seems like this is a weak spot, a soft target for terrorists. They can send a bird up, you know, and knock out the whole system.

This is one of our major international ports. We are Pacific rim, and I am very concerned about whether it reaches over to the FAA and if the FAA will look at all of our airports. Because it seems to me on September 11 it was—the airport was the scene—the launching of a terrible disaster that we've never had before and we were not prepared for. So in looking at how we prepare I think something like this should be a function of the FAA, and I would hope that Homeland Security would certainly raise these issues and see if we can motivate and activate FAA to take a look.

Mr. BROWN. Yes, ma'am. I'll certainly pass that information and the story along.

Ms. WATSON. I'll give you a copy of this, if you would like.

Mr. BROWN. Right. I'll pass that on to Under Secretary Libutti of the Information Analysis and Infrastructure Protection Directorate in the Department, because they are taking a significant review of all the critical infrastructure in this country and how we can better protect those vulnerabilities.

Ms. WATSON. Sure. Thank you very much.

Chairman TOM DAVIS. Ms. Watson, thank you very much.

Mrs. Maloney, any questions?

Mrs. MALONEY. No.

Chairman TOM DAVIS. Mr. Brown, thank you very much for being here today. We look forward to continuing to work for you as we develop these COOP plans. I will get my pronunciations right, and we'll get you armed with some funds when you come back here. But we could use—you know, we will look forward to working with you as we continue to develop these.

I'd just ask one last question. There is some concern among Members that maybe we ought to have these plans given to this committee where we could oversee them when they come in as well. Do you have any objection to that? We don't have to do that legislatively necessarily, but, as you get the plan, share them with us so we can stay abreast with what's going on.

Mr. BROWN. We will certainly continue those discussions, Mr. Chairman, and see if there isn't some way that we can have you more attuned to what we're doing in terms of planning and the processes.

Chairman THOMAS. Again, this may—hopefully, this will be—we're talking about events that never happen. We are talking about plans that never need to be implemented. But should they do that, all eyes will be on what we were doing in Congress.

Mr. BROWN. I would be remiss if I didn't remind ourselves that these COOP plans really go beyond just terrorist events. We also prepared to COOP the executive branch during Hurricane Isabel. There are many natural hazards which will cause us to COOP also, not just a terrorist event.

Chairman TOM DAVIS. Plus the tractor man. Isolated incidents.

Mr. BROWN. Right.

Chairman TOM DAVIS. And we had another guy on the bridge—just so I can get this off my chest—who was having a bad day and held up traffic on the Woodrow Wilson bridge and clogged traffic on the east coast for 5 hours. It took them 5 hours to figure out—they talked him down, instead of shooting him off with a bean bag, which is what they should have done right away, I mean, because you have to look at the greater good of some of this. It wouldn't have killed him, you know. He would have gotten wet.

But these kinds of plans sometimes we don't think about till they occur, and now that we have this agency we are expecting all knowledge to rest with you all and solutions to rest with you.

Mr. BROWN. We take this very seriously, and we will do everything we can to move it.

Chairman TOM DAVIS. We think you're doing great. Thanks for being here.

Mr. BROWN. Thank you, Mr. Chairman.

Chairman TOM DAVIS. We'll take a second and move to our third witness, and we have John Kern from AT&T. He's the Network Continuity Director for AT&T. This is a real-life company that has to deal with these kind of issues every day. This is part of their business, is dealing with emergency contingencies and service.

Mr. Kern, if you'd rise with me. It's the policy of this committee. We swear in.

[Witness sworn.]

Chairman TOM DAVIS. Please have a seat.

Your total testimony is in the record. We try to keep the opening statements to 5 minutes, but if you need to take a little longer, we are not in a hurry here. After 4 minutes, an orange light will go on. That gives you a minute. And when the red light goes on, that is it. Take what you need.

Thank you for being with us. I think you can add a lot to our testimony today.

STATEMENT OF JOHN KERN, DIRECTOR, NETWORK CONTINUITY, AT&T CORP.

Mr. KERN. My pleasure. Thank you.

Good morning, Mr. Chairman and members of the committee. My name is John Kern. I am the Network Continuity Director for AT&T. My team and I are responsible for business continuity, disaster recovery and continuity of operation for our worldwide network infrastructure.

Thank you for the opportunity to discuss with you today how AT&T has implemented our continuity of operations plan. I will suggest recommendations of how Federal agencies can implement continuity, plans of their own that kind of fall in line with some of the processes that we use.

The chart that is being displayed right now is an example of our network continuity and business continuity program, which is very similar to the COOP. We understand how important the services that we provide to our customers, both the private sector and the Federal Government services, like the government emergency telecommunications services; and we spend a great deal of energy and commitment to making sure that they can operate under any circumstances. This is both for our physical network and for cyber issues like security.

For example, we have basic level fire walls, intrusion detection at a higher level, cyber security where things are detected automatically and there is basic patterns that are looked for in the network so we can protect our services for our customers.

At a physical level, we have a dedicated team of people and we have invested over $300 million in equipment to be able to operate our network and continue our network under any circumstances. It is unique in our industry. We have had 12 years of experience and expertise in developing this as part of our continuity of operations plan.

The next one. One of the important things that was discussed today is exercising. I agree with the former witness that any plan that isn't tested really isn't a viable plan, and the whole point of an exercise is to find areas to improve the plan, to understand what can be done better the next time and how to make the continuity operations plan a viable, executable plan.

We realize it is a long process to do continuity of operations and business continuity. It took a substantial effort and discipline on our part to get this far in our plans and commitment. We have been working with the GSA to provide agencies with multiple suites of security services, and we look forward to continuing to work with the GSA and your committee to bring continuity of operations planning across the Federal Government.

It is obvious that continuity of operations planning is hard work. It requires investment. There is a cost to do it. In some cases, though, there is a larger cost in not doing it. Not having a continuity of operation plan that you could execute could mean that for several days your agency or enterprise isn't able to provide the basic service to your customers or your constituents.

The government should consider leveraging capabilities that have already been implemented in the industry, leveraging off the expertise of AT&T. The government business is very important, both to our customers, to the constituents of the government, and we are basically here to help with your continuity of operations plans. We have had a lot of benefit from our relationship with the Federal Government, various agencies, Department of Defense, FEMA, National Institute of Science and Technology for standards. We are now kind of offering what we have leveraged into those continuity of operations plans to offer assistance to you, Mr. Chairman, and to the committee and to the Federal Government.

Chairman TOM DAVIS. Thank you very much.

[The prepared statement of Mr. Kern follows:]

Statement of John Kern
Director of Network Continuity
AT&T
Written Testimony Given Before
The House Committee on Government Reform
April 22, 2004

Good morning, Mr. Chairman. My name is John Kern, Director of Network Continuity for AT&T. Thank you for the opportunity to testify today about how AT&T implements its own Continuity of Operations Plan and how we provide COOP services to our customers. As you know, AT&T provides premier telecommunications including network services, information systems and professional services to the government and to the private sector. I will share with you our recommendations regarding how Federal agencies can cost-effectively obtain COOP-supporting services from the commercial sector.

AT&T provides resilient connectivity, hosting and application services to meet or exceed our customers' business needs and continuity requirements. As the nation's largest Internet backbone provider, interexchange carrier, non-Regional Bell Operating Company Local Exchange Provider, and provider of network services to businesses, we are routinely challenged to engineer and operate a network of unparalleled scale. Like our colleagues in the industry, the National Communications System has tasked us since the peak of the Cold War to provide a variety of National Security/Emergency Preparedness (NS/EP) services built to meet very unique requirements. Some of these requirements have resulted in COOP initiatives to help ensure that the AT&T network is resilient and survives, thus ensuring COOP functionality for the government. We regularly exercise these capabilities and our proprietary disaster recovery strategies that are unique and unparalleled in the industry. We exercise under a simulated, and unfortunately, sometimes a real incident environment, as evidenced by our response to the terrorist attacks of 9/11. Our response and recovery capabilities enabled us to be responsive to the needs of maintaining critical telecommunications services for the Nation as well as our most demanding customers. We continually work closely with customers to accommodate their increasing business continuity needs – such as those recently defined by legislation.

Providing resilient COOP services for our large portfolio of customers begins with having integrated continuity capabilities and built in security best practices. We have built these

Statement of John Kern
Director of Network Continuity
AT&T
Written Testimony Given Before
The House Committee on Government Reform
April 22, 2004

capabilities into an organizational framework that facilitates their execution, both during quiet times and times of great stress. We established and utilize a rigid methodology of governance, risk assessment, protection, attack detection, and response. From the Chairman on down, every manager and organization is appropriately tasked for successful program execution.

AT&T established a system level Certification and Assurance governance process whereby we measure our estimated likelihood of recovery in the event of an incident. We then drill down to the component level and assess the consequences of a potential failure and the impact to our business. We work to mitigate the risk of failure by either eliminating the threat and the vulnerability, or mitigation of the exposure. This process constitutes our rigorous business case analysis and brings clarity to investment decisions. We have broken down COOP activities into manageable components, such as physical and logical. We assess these components both for ourselves and on behalf of our customers.

Physical level COOP has many facets. It begins by having diversity of communications links and equipment. When links and associated systems fail, there must be instantaneous and seamless rollover to backup facilities. This capability must be periodically tested, and given the frequency of cable dig-ups throughout the country, this testing occurs frequently. Service restoration must begin immediately after a disaster occurs. Our unique Network Disaster Recovery capability, the product of a $300M+ investment over the last ten years, enables us to replicate all of the components housed in our switching centers so that we can recover full functionality within 72 hours of destruction. To my knowledge, AT&T is the only telecommunications provider with this capability. Over 150 trailers stored in multiple locations around he country are staged to be deployed at a moment's notice. We exercise this capability at least four times per year by deploying the equipment and trained personnel to the designated recovery site. If you will let me know of your interest, I would welcome the opportunity to host your visit to one of our exercises. We will be in Miami next month, as well as San Francisco and Minneapolis later this year.

The next level of resiliency, the logical level, is where we sustain and protect our computer network facilities and interconnecting systems. We use a combination of firewalls and intrusion detection sensors and systems to ward off potential attackers. As you know, the sophistication and frequency of worms, viruses and Trojan horses is steadily rising. We are combating this phenomenon by anticipating attacks and executing strategies to defend our self and our customers. Then, rather than take it for granted that we've stopped all malevolent traffic at the perimeter, we perform similar functions within our network borders. We proactively analyze the traffic as it crosses our network to detect anomalies that would signal abnormal behavior. We essentially send a copy of all the traffic to a central computing facility that analyzes all the traffic flows for possible virus and worm signatures. When we find something that does not fit our normal signature patterns through our alarm systems, we perform forensics and may determine the traffic is a potential cyber threat. In such cases, a filter is created and deployed for network protection. Viruses and worms do not instantaneously appear but are developed over time and tested slowly and carefully so as not to attract notice. We have created the ability to evaluate the signature of customer data packets across our Internet backbone, detect attacks, and create filters for our network and information system assets and for those of our customers. As the level of sophistication steadily increases, we must provide updated tools and knowledge for continued increased vigilance to the detection of new viruses, worms, and Trojans.

Customers that host their applications with AT&T, or choose AT&T-provided applications, are routinely provided these protections and more. The same technology that we use for scanning our systems for security breaches is applied to our customers' host systems. We've also expanded our disaster recovery offering capability to include the back-up of customer data centers.

Statement of John Kern
Director of Network Continuity
AT&T
Written Testimony Given Before
The House Committee on Government Reform
April 22, 2004

In the Federal government markets we've been working closely with the GSA to provide agencies a portfolio of security services. Government missions are too important to not have COOP. Fortunately, industry now has the ability to be responsive to these needs. After extensively surveying agency security needs, with industry they developed a Multi Tiered Security Profile that provides agencies 4 suites of security services to accommodate 4 generally accepted levels of needed security. FTS-2001 contracts, including AT&T's, were modified last year to provide this capability; and there is now traction in the uptake. For example, the recent award of the Justice Unified Telecommunications Network, or JUTNET, to AT&T included the GSA-sponsored profile.

Now, as GSA is working the next generation FTS, called FTS Networx, we know from their industry surveys that GSA will be asking industry to provide varying levels of resilient services, defined by Service Level Agreement, to accommodate an agency's specific requirements. Through this open dialog we're confident that GSA has the understanding of the options, both commercial and non-commercial, so that FTS Networx will provide agencies a responsive suite of security products and services to accommodate COOP needs. As appropriate, AT&T will continue to work with the agencies, GSA and the Interagency Management Council, to assure the availability of our tested and scalable security solutions.

In summary, I would like to leave you with these three key points:

1) Industry faces many of the same COOP challenges that agencies face
2) Industry has developed solutions to these challenges for both internal use and for use by customers
3) The scale of industry network, IT security investment, and capability should be leveraged by the government so that COOP can be affordably and timely acquired. AT&T's multibillion-dollar investment in security and business continuity in multiple levels of the business, and those of our colleague carriers, should be exploited by the government to the maximum degree possible to minimize overall government expenditures.

I thank you for your time and I am available to answer any questions that you may have.

AT&T COOP Overview

Presented by John E. Kern, Network Continuity Director

AT&T

The world's networking company℠

Network Disaster Recovery

AT&T BC/DR Best Practices
Process Overview

AT&T Global Networking Technology Services – Network Operations

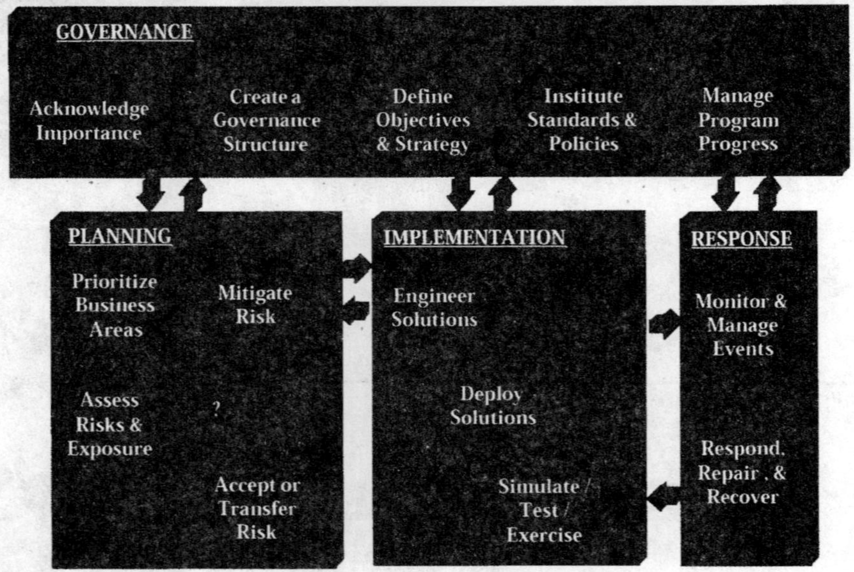

AT&T

Network Disaster Recovery

AT&T BC/DR Best Practices – Implementation
AT&T NDR - Balanced Equation

AT&T Global Networking Technology Services – Network Operations

NDR Operations Team

NDR DECT

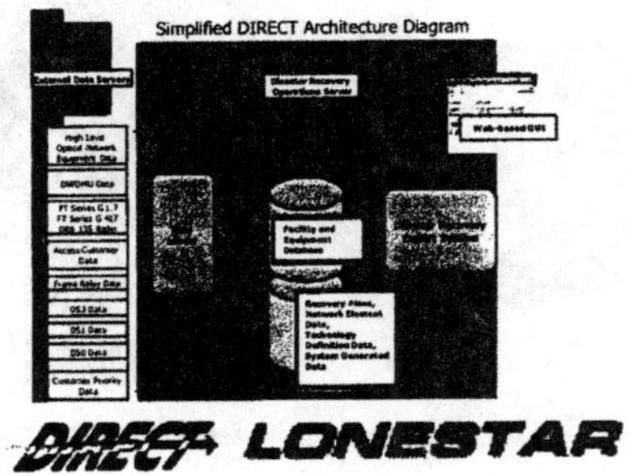

NDR Technology Trailers

AT&T

Network Disaster Recovery

AT&T BC/DR Best Practices – Implementation/Response
Test, Manage, and Recover

AT&T Global Networking Technology Services – Network Operations

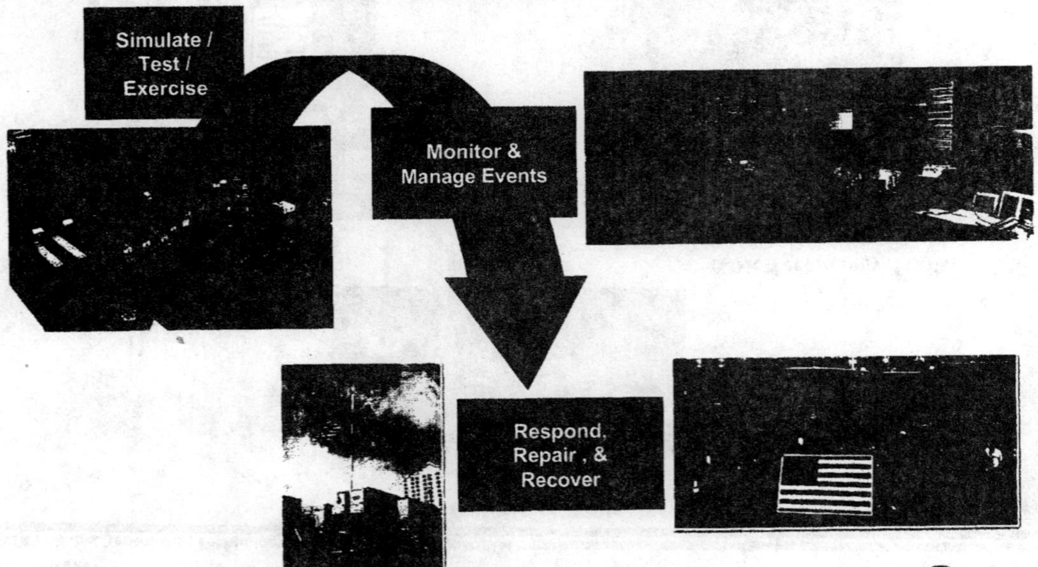

AT&T

Chairman TOM DAVIS. You get a lot of real-world experience in this. Every time there is a storm or something like that, you have to deal with that.

Mr. KERN. Yes, sir.

I'm an operational level person. If there is a disaster on my team, I go out in the field and do whatever we need to do to make sure our network continues to operate under any circumstances. We were heavily involved with our network recovery efforts after the World Trade Center.

One thing I mentioned a little bit earlier was having that discipline of a plan, the commitment to execute the plan and even having the resiliency and reliability built in. But you also need some flexibility in your plans. A good example, we had never and I don't know of anybody envisioning somebody crashing planes into a building the size of the World Trade Center or the subsequent shutdown of the nationwide air traffic control system. Our plans didn't call for that or didn't counter that. But the flexibility we built into our disaster recovery plans basically assumed that we would have regional disruptions.

A hurricane going through south Florida might shut down several airports. An earthquake in the West might shut down a few airports. So we have our people and our equipment regionally deployed so we can respond from anyplace. After the World Trade Center, when the air traffic system shut down, basically was a small inconvenience. We had people driving east to New York versus normally getting on a plane.

Chairman TOM DAVIS. You have a lot of redundancy in your system, don't you?

Mr. KERN. One of the things we do is we believe there is that kind of continuity by design, not just assuming what is going to happen in a disaster but how do you build that reliability and resiliency into your network or the service or infrastructure you need to provide your services for your customers or constituents.

For example, just in providing power, I know I mentioned the power outage at LAX. As far AT&T's offices were, we create the communications that basically hub the transport for our customers. We have three or four different levels of reliability around power. We have separate power feeds from separate substations. So, hopefully, the bird spreading its wings across one power line wouldn't impact the other power line.

We have dedicated generators in each building. We have battery backup, and we have the ability to bring in portable generators, kind of multiple layers of reliability and resiliency. I would say that a power outage would never be noticed by our customers because it is something we have built into the system. We wouldn't have to recover from it. We have planned for it and have built it into the network itself.

Chairman TOM DAVIS. One of the reasons we got you here today is because you know how to do this business. You do it on a practical basis. You are culturally a lot different than government. You have a lot of real-world experience in this at AT&T. Every time there is a massive storm, who knows, whatever disaster. So government is dealing with theoretical exercises. You are dealing with real-world experience; and nothing beats experience, as you know.

They used to say the difference between education and experience is education is when you read the fine print and experience is when you don't. In this particular case, you get your mistakes out already because you have a lot of experience with that. The government doesn't.

So, second, you are in a competitive atmosphere. These are not theoretical occurrences to you. These are occurrences that if they happen and you can't satisfy your customers, they can go to a competitor. In government's case, they have nowhere else to go. It makes you respond differently.

If government would try to take some of those competitive spirits you have—and try and tell the government and say this is why you operate it differently or this is why you make it more of a priority than government does. We try to do it in government sometimes, but these agency heads, they have a lot of pressure on them to perform under a lot of different regulatory obligations and this is when it probably won't happen, at least on their watch. You tend to push it aside, to put your resources toward something that is a little more current and a little more mission critical.

Mr. KERN. One of the things we had done to get past that—because in the early days of our program we had similar issues where the different organizational heads said I have more important things to do. This type of disaster will never impact us. One thing that we have set up as part of the process was kind of the governance structure. What's the set of standards and rules around what every organization in the government, what every agency has to do?

Probably most importantly and one of the functions that we perform that definitely would be a good idea for the government is in our case it would be a business impact analysis. In the case of the government, it might be an operations impact analysis. Understand that across the entire enterprise and government, what agencies are responsible working together to provide certain key services and functions to the constituents and then how do you address continuity of operations based on those critical services, not just on an agency level.

The other piece we have introduced over the 12 years that we've been doing this is the idea that this is part of a person's function, this is part of their job. For us in private enterprise, it goes to—the future funding they receive goes to their pay, future promotions; and it is in a sense of how they are graded. It's another important piece as a common report card.

If you have checklists, it is one thing. The next layer down is to look at what is the report card so that you know that a level A from one agency means the same as grade A for another agency and you do the exercises that the gentleman from FEMA mentioned that are across multiple agencies, kind of driven to a specific service.

Chairman TOM DAVIS. That is an excellent point.

Kind of mystifies me when you have an intelligence failure in the government, nobody gets fired. You have it at AT&T. You have a lot of people losing their jobs.

I am not arguing one is necessarily better than the other, but we could use a little more of the AT&T culture sometimes in government in staying ahead of the curve. But because we are not in a competitive mode, we tend to be more reactive than proactive.

You talk about incentives for your managers. There is no incentive for getting this plan down and having a great contingency plan. They are going to care more about current operations and what are you doing currently. I think that is the point you are making.

You also have to deal with a lot of changing technologies in telecommunication at this point, the move to wireless. Instead of interagencies, you have to work with your competitors in some cases, your line-sharing. How does that work?

Mr. KERN. The one issue is changes in technology, in some cases, changes in technology presents a challenge and some cases it presents a new opportunity. If you look at the increase in wireless technology, in the past, if you had to go to a physical place to connect into the network to get your job done or to get your business accomplished, now you can accomplish it wirelessly. In some cases, technology presents a challenge, but in a lot of our cases, it just presents more of an opportunity.

In the case of government, there is just more opportunity to leverage what is already being developed in the private industry to do a good COOP plan. If you imagine the wireless lands and wireless cellular voice technology, it allows you to set up your continuity of operations sites in places where you would not be able to get land lines to.

As far as the question around the cooperation, one of the things that we do through the Department of Homeland Security there is a National Coordinating Center, and that is one place in the industry where we can work together in the event of a real disaster or an event that would impact the network, like the power outage last summer, where we can get together to coordinate our activities to make sure if there is any mutual aid that makes sense where can we offer assistance where another carrier might not have enough generators or enough manpower to get a certain task done.

What places in the Federal Government offer that place of coordination and command and control that the telecommunications carriers get through the National Coordinating Center is another question that kind of brings to FEMA's role or not.

Chairman TOM DAVIS. Has anybody from the government come and asked you, what do you do for your COOP planning? Anybody consult with you and say you guys have to go through this? You have been through a lot of natural disasters and the like.

Mr. KERN. Personally, I have had dealings with several different agencies about their COOP plans, either reviewing them, offering suggestions on things that could be done or, in some cases, we will receive requests from government agencies to understand how the services that we provide, something like an ultra available, which is a way we can distribute technology across a given metropolitan area to make sure you don't have a point-to-point facility that is going to impact your ability to operate your enterprises, this kind of gives you a ring of capability, a place where you can operate your different services. So we will have requests from agencies to provide technology or to provide capability that they can use in their COOP.

We have had all different flavors. I think the agency we have dealt with the most has been the GSA where, again, part FTS 2001, there was a whole level of specific security applications that ranged for different levels of security that agencies could use to implement security needs.

We are also working with the GSA on FTS Networx, which is the next evolution of how do we not build in just the security but the resiliency and reliability. Each agency does not have the same need for the robustness, reliability, resiliency. How do you have a four, five-tiered structure so that agencies can get the reliability that they need to buy the resiliency to allow them to operate their business without agencies that don't need that same level of resiliency having in a sense pay for a service they are not going to use.

Chairman TOM DAVIS. Thank you very much.

Ms. Watson.

Ms. WATSON. I want to sincerely thank you for being here, Mr. Kern.

I would hope that your researchers can look toward the future. Everyone is saying, who would have ever thought an airplane would hit a building? We heard it rumored around before September 11. Now we know it is a reality. We need to look toward the future with our technology. We just put an apparatus on Mars, and they plan for it to go over rough surfaces and pick things up and photograph things.

What I am hearing that frustrates me is that we are not thinking progressively enough. I am very frustrated that we have had our third outage, as I mentioned, in 10 days. Why are we still depending on wires that go above the ground if a bird can light on them and knock out the whole airport? Are we thinking about the possibilities? We don't want a play on words, as was raised with the last panel. We want to really get people out ahead of these occurrences.

I think you could be very helpful, AT&T, in saying to our agencies, look, we have a design here that might work so you won't have to have this happen again; and then it is their responsibility to take a look and investigate. I would hope that you—and I know the competition is high, but come out ahead of all the others with a way to avoid—and I think that power outage can be avoided if we think more progressively and more scientifically. Maybe we ought to contact NASA, because apparently they have plans for all contingencies when they put a spacecraft up. But I want to encourage you to impact on us in government.

And I think the chairman was absolutely right. You know, we don't have the experience, and we don't get into the business of detecting things before they happen. We have not been in the business of doing that. We can make policy afterwards.

But I do think we are going to have to go out to the utilities, go out to private industry and say help and present to us, to FEMA, to the COOP or COOP or whatever you want to call it, you know, these are some things that government ought to invest in.

So I want to thank you for coming, Mr. Kern. I really don't have a question. It is more or less a recommendation to you to come back to us.

Chairman TOM DAVIS. Ms. Watson, let me just say, to stick with the puns and keep them on subject with the birds and the wires, you can't do this on the cheep.

Mr. KERN. My opening remarks mentioned my extra kind of capacity—I have seven acres in New Jersey. I have about nine hens and a rooster, so I am familiar with coops both at a business level and a personal level. But now that you throw cheep into the bargain—we are definitely linked to assist the government wherever we can. We have—over the 100 years that we have been around as an enterprise, AT&T has developed a very comprehensive set of standards around things like physical infrastructure. How do you power an enterprise that is important to you? How do you back that power up? What do you do around cyber security, physical security? How do you have continuity of operation plans that really take into effect where you can bring your people to—impacts to things like telecommuting, all the things we have great expertise at and definitely willing to help the government wherever we can either through our technology, our standards, our expertise or the experience that we have really developed over, in some cases, the last 12 years for business continuity but, in other cases, 100 years in operating a rather large, a rather critical infrastructure that provides the network service that everybody relies on.

Ms. WATSON. Mr. Chairman, if I could just take 1 more minute. We are going to have a bill on the floor, the Sensenbrenner bill. Reading the fine print—and this is where policymakers comes in. We read the fine print. We don't go on our experiences. It says that if there is an extraordinary circumstance and the Speaker of the House of Representatives announces vacancies, well, if the plane has succeeded in hitting the Capitol, it might have wiped everyone out, including the Speaker. If we are going to put law in the books, we are going to have to think beyond the words here. So it should be designated—someone who does the designating. Because the Speaker and all the rest of us will probably perish if that were to occur.

My point is we have to think differently than we have in the past; and, as a policymaker, this becomes the law. You know, it can be adjudicated in the courts. So how do we think in a way that will address these unusual circumstances?

Those of you out in the field in terms of the way agencies work and operations work and utilities work and so on have to benefit—we have to benefit from your experience, and you have to suggest to us. Now whether we make policy based on the input is left up to us, but I really invite your recommendations.

With that, I want to thank you, Mr. Chairman.

Chairman TOM DAVIS. Ms. Watson, thank you very much.

Mr. Ruppersberger, do you have any questions.

Mr. RUPPERSBERGER. I didn't hear your testimony. Thank you for being here.

Just generally, though, in the event that there is a catastrophe, it seems to me that in your field in communications it is an essential function during an event, after an event and then the months after the actual event. Do you communicate or work closely with anyone in Homeland Security as far as developing——

Mr. KERN. Yes.

Mr. RUPPERSBERGER. As much you can, what is that communication? Are they giving you the lead or are you helping them as consultants? How would you describe where that is now?

Again, we know Homeland Security is new. What would you like to see to make that function even better?

Mr. KERN. We have many roles with the Department of Homeland Security. One of them and probably important to me is the National Coordinating Center. It is the part of the Department of Homeland Security where the carriers have a common meeting ground to both plan around continuity and also to respond to an event.

During the World Trade Center, we worked through the NCC—at that time, it was part of the FCC—to understand where we need to bring in equipment or where we needed to have people. So the key function of the Department of Homeland Security for us is that kind of coordination role.

Another one is if we consider—what we try to do is not wait for the disaster but how do you get ahead and be proactive and look at the events that are coming up that you might need to worry about the impacts on your network or your people. I look at national security events as a big concern when one is declared by the government, understanding what is the real risk, what is the impact to our network. Do we need to do something different ahead of time to further harden our network, to bring in additional people in a nearby area? That is one area where the Department of Homeland Security definitely takes the lead around coordinating, around the contingency planning for national special security events; and we work through them.

Mr. RUPPERSBERGER. Are you coordinating your networks for—using your own money. Do you use Federal money? Where are you as it relates to money?

Mr. KERN. Right now, any of the planning that we do, any contingency planning that we do, it's our own money. As far as I know, we have not received any grants or funding to do our disaster recovery work or to do any of the contingency work. It is something that we have determined that is important to our customers, to the services that we provide and our ability to operate them under any event. We decided to undertake the expense and risk to do that.

Mr. RUPPERSBERGER. If an event occurs, another event in a major metropolitan area, are you ready?

Mr. KERN. Yes. We have had our program for the last 12 years. We were prepared for September 11, not for that type of event. But based on the structure and contingencies we had in place we were able to respond and deploy equipment to meet the needs from that disaster. Since September 11, we have increased our capabilities and added more people to the process and we are looking at things, some of the risks that are out there, maybe have a higher probability, the more manmade, chemical, biological attacks. We are participating in TOPOFF 3, which is the WMD exercise that is going to be held in New Jersey, Connecticut area next year. We have increased our capabilities to respond to those new threats. If there is an event in this country, we are prepared to respond.

Mr. RUPPERSBERGER. What about your other major competitors? Are they in the same position you're in, based on your knowledge? I know you are going to say you're the best, but are they close?

Mr. KERN. We don't spend any money looking to see what our competitors are doing with investment money.

Mr. RUPPERSBERGER. From a national security point of view, in the event there is a catastrophe, are we able to provide the communications needed? Because you are not the only game in town.

Mr. KERN. Unfortunately, that question would be best left to the competitors. I would say that none of our competitors have the mobile recovery capability that we have developed, the resiliency that we have developed, the multiple layers of backup that we have developed. To my knowledge, none of our competitors have taken their services as seriously as we have and do not have that type of capability.

We have invested more than $300 million. We have 150 pieces of mobile disaster recovery equipment dedicated to AT&T's network, both private enterprise and the Federal Government.

To my knowledge—I have been in the telecommunication industry for more than 28 years, and I have been in the disaster recovery field for more than 7 years, and none of our competitors have a mobile recovery capability to the extent that we do and could not respond in the same fashion that we can.

Chairman TOM DAVIS. Thank you very much.

I want to thank the Members for attending and thank our panelists.

Mr. Kern, thank you very much. This has been very helpful to us; and we wish you luck in your future endeavors as well.

Again, I want to thank our witnesses for attending. I would like to add that the record will be kept open for 2 weeks to allow witnesses to include any other information in the record.

The hearing is adjourned.

[Whereupon, at 11:30 a.m., the committee was adjourned.]

[NOTE.—The GAO report entitled, "Continuity of Operations, Improved Planning Needed to Ensure Delivery of Essential Government Services," may be found in committee files.]

[The prepared statement of Hon. Elijah E. Cummings and additional information submitted for the hearing record follow:]

Statement of Congressman Elijah E. Cummings
House Government Reform
Full Committee Hearing
On
"Can Federal Agencies Function in the Wake of a Disaster? A Status Report on Federal Agencies' Continuity of Operations Plans"
April 22, 2004 at 10:00 a.m.

Thank you, Mr. Chairman for holding this hearing to consider executive agency planning for the continuity of government operations in the event of a disaster, and the coordination role performed by the Federal Emergency Management Agency (FEMA).

In October of 1998, President Bill Clinton issued a Presidential Decision Directive (PDD) 67, to ensure the survival of a constitutional form of government and the continuity of essential federal government functions in the event of an emergency. At the time, he had no idea of the terrorist attacks that would occur on September 11, 2001, just three years later. Now, more than ever, with the increasing threat of organized terrorist activity, it is important that our government agencies are able to coordinate continuity of operations (COOP) in the wake of any calamity that could disrupt normal operations.

We come here today, to review the February 2004, report issued by the United States Government Accounting Office (GAO), titled, "Continuity of Operations: Improved Planning Needed to Ensure Delivery of Essential Government Services." The title alone does not instill much confidence in the reader, and the actual unclassified report findings identify deficiencies in both agency planning and Federal Emergency Management Agency (FEMA) oversight, further eroding the assurance that our agencies are prepared for a disaster. The Under Secretary for Emergency Preparedness has himself agreed that better COOP planning is necessary and that FEMA should do more to ensure the delivery of vital services.

Some of the many concerns noted in the GAO report were:

- Lack of uniformity in identifying essential functions of the agencies;
- Failure of many agencies to even develop plans and procedures for the performance of essential functions;.
- The lack of many agencies to develop clear procedures for adopting succession of leadership;
- And, even agencies inadequate conduction of tests, training, and exercises to ensure success in the event of an emergency.

Our government must take these findings seriously. Agencies **must** develop effective and adequate procedures for the continuity of government operations in the event of an emergency.

I look forward to hearing from the witnesses we have before us today. I hope that they are able not only to give an adequate assessment of the current COOP plans of individual agencies, but also to outline what steps should and will be taken to ensure that these agencies' COOP plans are satisfactory.

Thank you Mr. Chairman for holding this hearing.